An Introduction to
Tropical Grassland Husbandry

An Introduction to Tropical Grassland Husbandry

R. J. McILROY

SECOND EDITION

OXFORD UNIVERSITY PRESS

1972

*Oxford University Press, Ely House, London W.*1

GLASGOW NEW YORK TORONTO MELBOURNE WELLINGTON
CAPE TOWN IBADAN NAIROBI DAR ES SALAAM LUSAKA ADDIS ABABA
DELHI BOMBAY CALCUTTA MADRAS KARACHI LAHORE DACCA
KUALA LUMPUR SINGAPORE HONG KONG TOKYO

FIRST EDITION 1964
SECOND EDITION 1972

Set in 10/12 pt. Monotpye Times New Roman, printed by letterpress, and bound in Great Britain at The Pitman Press, Bath

Foreword to the First Edition

WILLIAM DAVIES
Former Director of The Grassland Research Institute, Hurley, England

Professor McIlroy's book is most timely for it goes a long way to cater for the needs of the student of tropical pastures who, up to the present, has not had adequate teaching in the basic principles of pasture/grazing animal husbandry in the tropical environment. Hitherto the emphasis has been upon the production of bread corn and the plantation crops but, important as these are, it seems quite inappropriate not to place at least equal emphasis upon the teaching of agronomic principles concerning grassland and grazing animals.

It is well to recognize the important part which pastures and other fodder crops play throughout tropical agriculture. Within the tropics there are probably 10 000 million acres of land to which grazing animals have access and which provide the main source for their sustenance. The overall contribution of these 'grasslands' is to sustain about half the domestic animals of this world and to produce about one-third of the meat and one-fifth of the milk products grown globally. These are material contributions from areas whose potential is not developed and, indeed, as yet is largely undefined.

Professor McIlroy indicates clearly that the application of the basic principles of plant and animal production is little different whether in tropical or temperate climes. It seems unmistakable that the factors of pasture growth and its use by animals that have found elucidation as a consequence of temperate country researches, can also be used as a basis for teaching in the tropics. It is, however, most necessary to ensure that soundly-planned pasture experiments be carried out over the widest possible range of tropical conditions. To do so requires trained pasture agronomists, and where better to start such training than by the teaching of already-accepted principles and meanwhile to provide cogent examples from situations within the tropics.

There is already a wealth of somewhat empirical evidence dealing with the tropical pasture/grazing animal environment. This needs to be recognized for what it is, namely the result of careful observation and often of long experience, recorded under conditions which offered little facility for critical experimentation. These data must not be disregarded for they provide the starting point from which to initiate more detailed research programmes. They also offer the surety that high rewards will accrue as a consequence of further research. If such further enquiries are to be made then it is essential that universities and agricultural colleges in the tropics should give their students sound teaching in the principles of grassland agronomy and in the application of the biological sciences to the growth and use of pasture. Here lies the great significance and timeliness of this book. It can be used with confidence by both teachers and students alike as a sound guide to their pastoral studies.

WILLIAM DAVIES

Preface

The present volume is based on a series of lectures on grassland husbandry given to undergraduates of the University of Ibadan, Nigeria, as part of the course for the degree of B.Sc.(Agric.).

The science and practice of grassland farming in humid temperate regions have reached high levels and it is not unreasonable to assume that a high standard of efficiency may be maintained with the application of existing knowledge. This is not true of tropical regions where much basic knowledge remains to be discovered. While the principles of grassland farming are to a degree universal there is a vast difference between the intensive grassland farming practised in the advanced countries of humid temperate regions and the extensive grazing practised by, say, the nomadic pastoralist of tropical Africa.

For generations the indigenous farmer of the tropics has derived his income from cash crops such as cocoa, coffee, oil palm, etc., after the immediate needs of his family have been met by the growing of food crops. Livestock were maintained for ceremonial and prestige purposes and for traction, although the sale of hides was not unimportant. It is only in very recent years that serious thought has been given by the tropical farmer to livestock products (meat, milk, etc.) as a source of income.

The development of livestock implies the development of grassland farming since grass is the cheapest food for the ruminant animal. The grassland potential of the tropics is enormous and, if properly developed and efficiently utilized, should provide animal protein to meet the requirements of not only the tropical population but of others besides. It is in the tropics that the greatest increase in production per acre can be expected. More than half of the cattle of the world are raised in the tropics but their plane of nutrition is woefully low.

Increased animal production in the tropics depends on

(1) overcoming religious and social prejudices against the culling of livestock;

(2) modification of existing farming systems;

(3) improvement in the quality of livestock;

(4) improvement in the quality and quantity of pasture and extension of the grazing season;

(5) the more efficient utilization of the pasture, involving the conservation of surplus feed produced in the rainy season, and the provision of supplementary fodder crops; and

(6) improved management of livestock and grazing pasture.

While there is an extensive literature on grassland farming of temperate regions, no introductory textbook is available to the student of grassland farming in the tropics. This volume attempts to meet this requirement. The general principles are outlined with especial reference to tropical conditions but where no tropical example is available recourse is, of necessity, made to temperate conditions on the assumption that is is better to discuss the principle in an unfamiliar environment than not to discuss it at all. The frequent references to temperate region agriculture further emphasize the present inadequacy of our knowledge of tropical grassland husbandry.

The author gratefully acknowledges the assistance given by Dr. William Davies, Mr. Geo. Jackson and Dr. D. H. Hill all of whom read the manuscript and made valuable suggestions which have been incorporated in the text. He is especially grateful to Dr. Davies for the Foreword and to Mr. Geo. Jackson for Appendixes A and B.

R. J. McIlroy

Ibadan
November 1962

PUBLISHER'S NOTE

It is with great regret that we record the death of Professor McIlroy shortly after he had completed the revision of this book. Our thanks are due to Professor E. J. Britten and Dr. P. J. Skerman for overseeing the proofs and revising the index.

Contents

1 The Economic Value of Grassland

No farmer is likely to extend or improve the grassland area on his holding or to include grass in his crop rotations until he is convinced of the economic value of grassland. The following table based on production in the United Kingdom demonstrates that grass, when grazed, is the cheapest food for ruminant livestock and that grass, when conserved as silage or hay, is cheaper than most other temperate-region feeding-stuffs.

TABLE 1

Estimated average costs of production of starch equivalent[1] (£ per 100 kg) *for various crops and dairy cake 1960–1.* (from *Costs and efficiency in milk production*, H.M.S.O., London, 1964)

Grazing	1·64
Kale (grazed)	1·70
Hay	1·79
Grass silage	2·05
Kale (carted)	2·56
Arable silage	3·19
Mangolds	3·39
Turnips and swedes	3·88
Barley	1·64
Oats	2·13
Dairy cake (S.E. 63)	4·78

[1] The Starch Equivalent (S.E.) of a feeding-stuff is the number of kilogrammes of starch equivalent in production of energy to 100 kg of feeding-stuff.

When farmed at its highest potential, grass yields more starch equivalent and protein equivalent per acre than any other crop. Yields of dry matter exceeding 11 000 kg per hectare per year may be obtained in humid temperate regions while yields of the order of 22 400 kg may be obtained in the tropics, e.g. the indigenous grasses *Chloris gayana* and *Cynodon plectostachyus* have produced 22 400 kg of dry matter per hectare per year in Malawi when heavily fertilized. Good grassland, effectively grazed, may have a starch equivalent of 66 per cent and a protein equivalent of 15 per cent. Good permanent pasture will produce 1200–3400 kg of total digestible nutrients (T.D.N.) per hectare. The output of the best feeding pastures in terms of live-weight increase per hectare per year may be as high as 1883 kg (or 1000 kg per hectare per year of meat) when intensively farmed, although a more usual figure for first-class permanent pasture in the United Kingdom would be 380–440 kg live weight increase (224–260 kg meat). These production figures are representative of the fattening pastures of the Midlands, Kent, and Northumberland in England.

In the tropics, large commercial units based on grassland farming are rare and markets as yet relatively undeveloped. High production costs are aggravated by low milk yields and late maturity. In the large cities, however, milk and dairy products fetch high prices and some commercial urban dairies have operated successfully for many years, e.g. at Bombay in India and at Khartoum in the Sudan. There is evident expansion of the market for meat in the urban areas of the tropics and, as production costs and consumer prices are reduced, the demand should become widespread. In Nigeria, with a population of about 60 million, milk is absent from the diet of about three-quarters of the population and, except in the cattle country of the extreme north and among the wealthier classes in the towns, there is a general shortage of meat.

The outlook for animal production would appear to justify the effort and expenditure required for the development of tropical grasslands and livestock. Unfortunately, few reliable figures are available on which to base an accurate assessment of the future of the industry.

2 The Nature of Grasses and Legumes

It is estimated that there are about 10 000 species of grasses (Gramineae) in the world. Of these, only about forty are used to any appreciable extent in the establishment of sown pastures. These cultivated pasture grasses form part of the indigenous flora of three main regions, viz,

(a) the Eurasian region with about twenty of the forty species;

(b) the East African region with eight species;

(c) a sub-tropical South American region with four species (Hartley and Williams 1956).

The majority of the species used in the cultivated pastures of tropical territories originated in regions (b) and (c).

The genera of grasses are distinguished from one another chiefly by the arrangement, form, and modification of the miniature leaf-like scales enveloping the flowers, while the species are usually separated by differences in duration (annual, biennial, perennial), form of growth, and in the size and shape of the stems, leaves, and the flower-head and its parts.

The general structure of grasses (after Armstrong 1937; Hubbard 1959)

The stem of grasses, bearing the leaves and flower-head, is termed the *culm*. It is cylindrical and often hollow except at the *nodes*, or joints, which are of solid tissue. The hollow portions between the nodes are the *internodes*. In a few grasses the basal internodes may be swollen or bulbous. The nodes form the points of attachment of the leaves, which are arranged in two rows alternating on opposite sides of the stem.

The upper, expanded part of the leaf is the blade or *lamina* and the basal part, which embraces the stem, is the *sheath*. The leaf blade often widens at the base into a ledge-like process, or may form ear-like projections on either side. The latter are termed *auricles*. At the point where the blade forms the sheath there is usually a membranous outgrowth called the *ligule*.

The production of new stems or shoots within the axils of older leaves is the normal method of branching in grasses and is called *tillering*. If the stems grow up within the leaf-sheaths of the older leaves dense tufts of foliage are produced (intra-vaginal mode of growth). If the new shoots pierce the sheaths near to their point of origin and grow out horizontally or obliquely (extra-vaginal mode of growth) loose or open tufts result.

Roots arise adventitiously from the lowest node or nodes of the stem and are thin and fibrous. In the case of certain perennial grasses, e.g. *Chloris gayana* and *Axonopus compressus*, the roots are produced at every node of a surface-creeping stem or stolon; in certain other grasses roots arise from the nodes of underground-creeping stems or *rhizomes*, e.g. *Pennisetum purpureum*.

The flower-head or inflorescence, which is terminal on the stem, is frequently a panicle consisting of clusters of spreading branches terminated by long structures called *spikelets*. These are composed of one or more flowers with enveloping scales that conceal the flowers from view except at flowering time. Alternatively the spikelets may be stalked directly on to the main axis, when the flower-head is a *raceme*, or borne on the main axis itself, when the flower-head is known as a *spike*.

The small flower, when bisexual, consists of one carpel and three or more stamens. The latter have long, slender filaments bearing long anthers. The ovary, which may be glabrous or hairy, bears two divergent feathery stigmas. At the base of the ovary there are two small scales termed *lodicules* whose principal function is to assist in the opening of the flower by pushing apart the protecting scales at sexual maturity of the flower. The reproductive organs (ovary and stamens) are enclosed between these boat-shaped scales. The upper or inner scale (*palea*) is usually thin and membranous; the lower or outer scale (*lemma*) is larger and stouter, overlaps the inner one, and frequently bears a bristle-like structure or *awn*. The lemma and palea are protected by two outer scales known as *glumes*. In a few genera, e.g. *Axonopus*, the lower glume, and in some both glumes, may be absent. The flowers of annual grasses are generally self-fertile while those of perennials are generally self-sterile and cross-pollinated by the agency of wind.

The naked fruit or grain of grasses is called a *caryopsis*. It is one-seeded with the pericarp fused to the testa (seed-coat) and consists of the embryo together with its store of starchy endosperm for nourishment of the developing seedling. The embryo consists of a primary shoot (*plumule*), primary root (*radicle*), and a flat, shield-like structure, the *scutellum*. The name 'grass-seed' is often applied not only to the naked fruit but also to the entire spikelet or even to a group of united spikelets.

Why grasses are suitable as herbage plants

Grasses are especially suitable as herbage plants for grazing by livestock or for mowing for the following reasons.

(a) Reproduction of fresh shoots by tillering provides a means of recovery from cutting or grazing.

(b) New tissue produced during growth arises chiefly at the base of the leaves where it is least likely to be damaged by cutting or grazing.

(c) Many grasses maintain continuous vegetative growth interrupted only by periods of drought or cold.

(d) Many grasses spread by rhizomes or stolons which readily form adventitious roots and give rapid ground coverage.

(e) The root system binds the soil particles together forming a 'sod' and brings to the surface layers nutrients that have been leached into the subsoil by heavy rainfall.

In Uganda, Martin (1944) has demonstrated the regeneration of soil structure and restoration of soil fertility that takes place when a resting period under elephant grass (*Pennisetum purpureum*) follows a period of continuous cropping. A resting period of three years under grass appears to be necessary. *Panicum maximum*, *Brachiaria decumbens*, *Cynodon plectostachyus*, and *Paspalum notatum* were also effective in the restoration of fertility. Griffith (1949) showed that a three-year fallow under elephant grass in Uganda led to an increase of 673 kg of nitrogen per hectare in the top 30 cm of soil or 224 kg of nitrogen per hectare per year.

The general structure of legumes (after Cobley 1956)

The pasture legumes are members of the family Papilionaceae, a division of the Leguminosae that includes thousands of species. They are mainly herbaceous plants or shrubs with leaves mostly pinnate and stipulate. The pinnae themselves are often subtended by stipules. In some cases, the leaf or part of the leaf is modified into climbing tendrils. A common feature is a pulvinus at the base of the leaflets and petiole.

The inflorescence is a spike or raceme. The flowers are usually subtended by bracts which vary in size. The structure of the flower resembles a butterfly and thus gives the name to the family. The flower is zygomorphic and hermaphrodite with five sepals, partially fused, enclosing the base of the flower when mature. The corolla consists of five petals. The upper and largest of these is called the standard. Two smaller lateral petals are the wings, while two lower petals fused along the lower or both margins form the keel of the flower.

The adroecium consists of ten stamens, which may be united by their filaments (*monadelphous*), or there may be nine united and one free stamen (*diadelphous*). The superior ovary, surmounted by a bent style with broad stigmatic surface, consists of one carpel bearing a single row of ovules. It is surrounded by the staminal tube.

The flowers are either cross-fertilized or self-fertilized, the fruit produced being a legume or pod, splitting when mature along both dorsal and ventral sutures.

The seeds vary greatly in size, shape, and colour. They tend to have a prominent aril running part of the way round the edge of the seed. The seeds are non-endospermic.

All the legumes produce nodules on their roots in which bacterial colonies live in symbiosis with the plants. These bacteria are able to fix atmospheric nitrogen and thus maintain soil fertility, a property of major importance in the tropics where many soils are deficient in nitrogen. The bacteria that infect the roots and stimulate nodule development are all species of the genus *Rhizobium*. Each *Rhizobium* species is associated with a definite group of legumes. Some legumes, e.g. *Phaseolus atropurpureus*, nodulate freely in the field whether

Rhizobium is applied to the seed or not. A high proportion require inoculation for establishment, and some show marked variation in nodulating ability even when inoculated, e.g. *Desmodium intortum* and *Centrosema pubescens.*

The property of nitrogen fixation with consequent elevation of the nitrogen status of plant and soil, and the acceptability of legumes to livestock make them desirable components of pastures. The first introduction of a legume (red clover) into pasture in association with grasses is attributed to Sir Richard Weston in 1613.

3 The Agricultural Value of Herbage Species

In selection of a pasture grass or legume species the main qualities desired are productivity, palatability, nutritive value, and adaptation to the local soil and climatic conditions.

The *productivity* or yield depends on

(a) persistence, or the ability to survive and to spread by vegetative methods;

(b) aggressiveness, or the ability to survive the competition of other associated species;

(c) the ability to recover from hard grazing and trampling;

(d) drought-resistance and tolerance to cold;

(e) seasonal distribution of production;

(f) the capacity to produce a good yield of viable seed or to be established cheaply by vegetative propagation;

(g) soil fertility (especially nitrogen level);

(h) climate.

Palatability has been defined by Ivins (1952) as follows:

> Palatability is the sum of the factors which operate to determine whether, and to what degree, the food is attractive to the animal; it can thus be held to constitute the connecting link between grass and the grazing animal and is regarded by various authors to be of greater importance than nutritive value. Necessarily relative, it is influenced by such variables as the animal itself, stage of growth and development of the herbage, alternative foods and the management and manuring of the herbage.

It has been demonstrated frequently that cows prefer indigenous grasses to selected varieties of grass, despite the fact that the indigenous grasses may be lower in productivity and nutritive value. Ivins compared several temperate-region species in palatability trials and found that *Plantago* was more palatable to cows than the otherwise more desirable herbage species rye grass, cocksfoot, timothy, and white clover.

The palatability may be tested in a 'cow cafeteria'. This consists of groups of plots of a number of different herbage species. Cattle are allowed free access to the plots, and the time spent on each plot or the amount of forage removed by the animals provides an index of the relative palatability of each species. In palatability trials the effect of fertilizer, especially nitrogen, on the palatability of the herbage should not be overlooked. High dosage of nitrogen of the order of 500 kg per hectare reduces palatability. Rogler (1944) has demonstrated that maturity, grazing intensity, recovery rate, drought resistance, associated species in a mixture, previous feed, kind of stock, and individual differences in animals all affect palatability.

Assessment of the *nutritive value* of pasture species is based on determination of the chemical composition and digestibility. These will be discussed in Chapters 13 and 15. Estimation of the leaf–stem ratio is important as the nutritive value of the leaf is greater than that of the mature stem. A stemmy habit is often associated with high seed production and short life and may be advantageous in species grown for seeds or hay. The nutritive value is influenced by the stage of growth at the time of cutting or grazing, by environmental conditions, and by fertilizer application. Application of nitrogen increases the nitrogen content of herbage but, in so far as it influences rate of growth, slightly depresses the digestibility. It has been shown by Minson and his co-workers (1960) that the digestibility of grass species falls off rapidly at ear emergence.

Methods for evaluating pasture grasses have been concisely summarized by Burton (1951). A good method is to plant them in rows or plots in pure stands and to graze these by the class of livestock that will be the eventual consumer. When grazing is not practicable, mowing to simulate grazing is adopted. The trials should include at least two levels of soil fertility and two systems of management. The means of propagation, whether by seed or by vegatative means, and also the method of eradication or control should be studied before a new variety or species is established as a pasture. Herbage varieties show a wide range of genetic variability and may be selected with varying growth habits and responses to management and environment.

4 Some Grass and Legume Species of Tropical Regions

THE agronomic characters of some of the more important herbage species grown in the tropics are summarized below. For notes on further species, F.A.O. Agricultural Study no. 42 (Whyte *et al.* 1959) *Grasses in agriculture* and no. 21 (Whyte *et al.* 1953) *Legumes in agriculture* should be consulted.

GRASSES

Andropogon gayanus (Gamba)

A tufted perennial up to 3m high. Common in Northern Nigeria and in parts of Africa that have a long dry season. Introduced successfully into Brazil and tropical Australia. Drought-resistant. Produces an abundance of new shoots after burning. Prepared by planting of splits or by seed. Seed rate 30–60 kg per hectare of uncleaned seed.

Axonopus compressus (carpet grass)

A shallow-rooted perennial with short stolons. Flowering stems 30–60 cm high. Native of Central and South America and the West Indies. Introduced into West Africa, Malaya, and Indonesia. Grows best in moist soils. Dies back in the dry season. Established by seed or cuttings. Forms a dense turf and withstands close grazing but its productivity is low, unless heavily fertilized. Can suppress *Cynodon plectostachyus.*

Axonopus affinis

Used for permanent pastures in Florida, West Indies, Fiji, Malaya, Guyana, and Hawaii. Mixture with *Desmodium canum* successful in Hawaii. Seed rate 6–14 kg per hectare.

Beckeropsis uniseta

A tall, tufted perennial up to 2 m high with leaves up to 60 cm long. Widespread in Africa. Drought-resistant. High-yielding. Propagated by seed.

Bothriochloa insculpta (pinhole grass)

A creeping, stoloniferous perennial with aromatic leaves. Forms a close turf when heavily grazed. Promising for permanent grassland in the drier parts of East Africa. Outyields Rhodes grass (*Chloris gayana*).

Brachiaria brizantha (palisade grass, signal grass)

Tall, rhizomatous perennial. Native of tropical Africa. A leafy, upland grass. Drought-resistant. Reported to cause photosensitization in Australia. Productive and palatable. Propagation by seed or division of root stocks. Introduced sucessfully into Ceylon, it has now become the most important pasture grass in that country.

Brachiaria decumbens

A creeping, stoloniferous grass, native of tropical Africa. Makes good cover in the wetter areas. One of the best pastures in the Congo in a pure stand or in association with *Stylosanthes guyanensis.* Introduced successfully into tropical Australia and the West Indies.

Brachiaria mutica (para grass)

Creeping, stoloniferous perennial with stems 2m or more high. Native of tropical Africa and South America. Suitable for moist, lowland pastures. Excellent fodder but must be consumed young as it readily lignifies. Will not withstand heavy grazing. Grows well in association with *Centrosema* and *Trifolium* in Queensland. Seed rate 3–5 kg per hectare.

Brachiaria ruziziensis (Kennedy ruzi grass)

Perennial; growth habit similar to *B. mutica* but leafier and later seeding. More productive and palatable than *B. mutica* but susceptible to frost. Prefers fertile, well-drained soils; rainfall 100–150 cm. Seed rate 2–3 kg per hectare.

Cenchrus ciliaris (African foxtail, Buffel grass)

Perennial, 30–90 cm high. Forms spreading tufts. Native of tropical Africa, India, and Indonesia. Cultivated for permanent pastures in Africa and Queensland. Remains green in the dry season but mature stems become wiry. Eaten readily during dry season but not so readily during the rains. Productivity low. Seed rate 3 kg per hectare.

Chloris gayana (Rhodes grass)

Leafy perennial 30–120 cm or more in height. Spreads by stolons which root at the nodes giving the appearance of a tufted sward when mature. Native to tropical Africa but widely introduced throughout the tropics and sub-tropics. Not suited to areas with more than 125–150 cm of rainfall per year. Withstands grazing well but new stands should be allowed to flower and set seed before being grazed or mown. Experience at Samaru in Northern Nigeria suggests that Rhodes grass is not suited to very dry conditions. Yield in Malawi 5 tons dry matter per hectare when unfertilized and 27 tons per hectare fertilized. Optimum fertilizer 45–67 kg nitrogen per hectare. When sown in mixed stands Rhodes grass is quickly suppressed by more aggressive grasses, e.g. giant star grass (*Cynodon plectostachyus*). Seed production good. Yields of seed of up to 560 kg per hectare have been obtained. Easily established from seed. Seed rate 8–9 kg per hectare. Often sown under maize as a nurse crop.

Cynodon plectostachyus (giant star grass, Naivasha star grass)

Perennial, with rapidly growing stolons that quickly cover bare ground forming a dense turf. Up to 120 cm high. Native of East Africa but common throughout the tropics. Withstands grazing well and close grazing is essential for the maintenance of a good pasture. Responds to nitrogen fertilizer and a high nitrogen status is required if the grass is not to produce too many long stolons. Palatable. Aggressive. Yield of hay in Malawi 6·25 tons per hectare when unfertilized and up to 25 tons when heavily fertilized. Forms a promising pasture in association with *Centrosema*. Some varieties produce good seed crops in areas of low humidity but propagation generally by vegetative means.

Cynodon dactylon
(Bermuda grass, Bahama grass, star grass, Dhub grass)

Found throughout the tropics and sub-tropics. Under the name of Dhub grass is one of the most important pasture grasses in India. Establishment and management as for *C. plectostachyus*.

Dichanthium caricosum (Nadi blue grass, Antigua hay grass)

Tufted and stoloniferous. 30–68 cm high. Native of India and South-east Asia. Drought-resistant. One of the best pasture grasses in the Pacific Islands and the West Indies. Combines well with *Stylosanthes hamata*. Withstands close grazing. Propagated by seed or root divisions.

Digitaria decumbens (Pangola grass)

Perennial, with long rooting stolons. Forms an open turf. Native of South Africa. Introduced successfully in the south-east United States of America, Hawaii, and the West Indies, where it carried one steer to 0·27 hectare with 0·45 kg daily average live-weight increase. In south Florida, Pangola grass has produced more beef than any other grass tested. Makes best growth on well-drained moist soils but will survive both flooding and drought. Persists under continuous close grazing. Propagated by planting sprigs or mowings. These may be broadcast and disced in at the rate of 1·25–2·5 ton of sprigs per hectare. The pasture should be topped by mowing or light grazing 6–8 weeks after planting. Pangola grass, especially in the Caribbean, has proved to be susceptible to attack by insect pests and a virus disease. Resistant varieties are being sought.

Echinochloa pyramidalis (antelope grass)

Perennial. Rhizomatous. Found throughout tropical Africa. A water-loving grass, valuable as fodder. Vigorous and leafy under irrigation. In Malawi it yielded 15 tons of hay per hectare when cut in April. Good seed producer but generally propagated by division of rhizomes.

Eragrostis curvula (weeping love grass)

Bunch grass, up to 60 cm high, Annual and short-lived perennial. Native of South Africa and Rhodesia. Drought-resistant. Low producer but prized as hay for horses. Used in South Africa for summer grazing and leys with lucerne. Winter pasture in Florida. Does not persist under continuous grazing. Good seed production. Seed rate 1 kg per acre.

Eriochloa polystachyus (Carib grass)

Trailing perennial. Native of tropical South and Central America and West Indies. Adapted to humid tropics. Palatable grazing and hay. Propagated by stem cuttings or division of root stocks.

Hyparrhenia rufa (Jaragua grass)

Tufted perennial. Up to 3 m high. Drought-resistant. Found throughout tropical Africa, Brazil, and Central and South America. Comes next to Gamba in importance as an indigenous grazing grass in Nigeria. Young growth valuable but of low nutritive value when in flower. Does not persist under close grazing. Established by seed or by division of root stocks.

Ischaemum aristatum (Batiki blue grass)

Perennial, with erect or spreading shoots. Native of tropical Africa and south-east Asia. Permanent pasture in humid tropics, in Fiji, Trinidad, and India. Poor seeder. Propagated by division of root stocks. The Trinidad variety is reported to taint milk.

Melinis minutiflora (molasses grass)

Perennial. Up to 1·5 m high. Widely distributed throughout the tropics and sub-tropics. Forms large tussocks. Sticky exudation on leaves and strong, sweet smell. A valuable grass. Fairly drought-resistant—will persist through 4–5 months of dry season. Must be well esablished before grazing. Combines well with legumes, e.g. *Pueraria phaseoloides*, *Stylosanthes guyanensis*, and *Phaseolus lathyroides*. Good seed production of up to 280 kg per hectare. Productive

Yields of up to 11·25 tons per hectare have been obtained. Not very palatable. Sensitive to frost. Established by seed. Seed rate in Queensland 2–4·5 kg per hectare; in Brazil 22·4–28 kg per hectare.

Panicum maximum (Guinea grass)

Tufted perennial. Up to 3m high. Native of tropical and sub-tropical Africa and found throughout the humid tropics and sub-tropics of the world from sea level to 2000 m. Does not form a sward but remains tufted. Combines well with tropical legumes. Does well with molasses grass and with star grass. Drought-resistant. High nutritive value when young. Palatable. Not so coarse as elephant grass (*Pennisetum purpureum*). Dies rapidly under continuous grazing. Yields of 122 tons green fodder per hectare per year reported from the Philippines and Australia but 32·5–40 tons reported from Malaya. Seeds readily but viability is low. Established by seed or divided stools. Seed rate 4·5–11 kg per hectare.

P. maximum var. *trichoglume* (green panic)

More palatable and drought-resistant than Guinea grass. Can be sown in most areas with 56 cm of rain or more. Responds to fertilizer and mixes well with legumes One of the most important pasture grasses in Queensland.

Panicum antidotale (blue panic)

Sown in areas that are too dry for Guinea grass or green panic; 50–75 cm rainfall.

Paspalum notatum (Bahia grass)

Deep-rooted, rhizomatous perennial. Up to 60 cm or more in height. Native of Central and South America and adapted to tropics and sub-tropics. A useful pasture grass, persistent to grazing. Fairly drought-resistant but dies out in the dry season in Northern Nigeria. Palatability generally considered to be low. Readily forms a dense sward and can be grazed 3 months after planting. Unequalled as a grass for soil conservation. Established by cuttings or seed. Seed rate 10–20 kg per hectare.

Paspalum coloratum

Very palatable and in Malawi has carried 20 sheep per hectare for 192 days. Very drought resistant. Seed rate 2 kg per hectare.

Paspalum scrobiculatum and *Paspalum dilatatum*

Important pasture and hay grasses in sub-tropical Australia.

Paspalum plicatulum.

A hardy grass, which can do well under fairly low fertility. Good drought-tolerance and can also stand waterlogging. Requires 76 cm or more rainfall. Valuable stand-over winter feed even when frosted. Seed rate 3–4·5 kg/ha.

Pennisetum purpureum (elephant grass, Napier grass)

Tall, tufted perennial. Up to 4·5 m high. Distributed throughout tropical Africa. Occurs naturally near to streams and rivers. Indigenous and distributed widely in Nigeria. Palatable, drought-resistant, and high yielding. Yields of over 180 tons of green fodder per hectare per year obtained under humid conditions or under irrigation. High nutritive value. Used in Uganda for restoration of soil fertility; 3 years under *P. purpureum* after 3 years' cropping led to marked improvement. In the West Indies *P. purpureum* persisted for 6 years under grazing carrying 5 steers per hectare from May to July and one steer to 0·24 hectare for the rest of the grazing period. An excellent grass for silage, the aftermath giving dry-season pasture if not grazed too heavily. Mixed stands with *Centrosema pubescens*, *Glycine wightü* *Clitoria* spp., or *Cajanus cajan* used for resting leys in Malawi. Established from shoots.

Pennisetum clandestinum (Kikuyu grass)

Perennial. Important in the cooler tropics, e.g. Kenya. Reputed to be the best pasture grass above 1500 m. Suffers in periods of drought. Established from seed or cuttings.

Pennisetum pedicellatum (Kyasuwa grass)

Annual. Established from seed. Gives excellent yields of hay or green fodder in West Africa and India.

Setaria sphacelata (golden timothy)

Tufted perennial up to 150 cm high. Distributed throughout tropical Africa. Persistent, palatable, productive, and drought-resistant. Long vegetative cycle. High nutritive value. Under best conditions crude protein may exceed 18 per cent with only 25 per cent crude fibre. Seed setting poor. Established from seed or cuttings. Seed rate 4·5–11 kg per hectare; in Queensland 2–3 kg per hectare.

Setaria splendida

Similar to but larger than *S. sphacelata*. Established from division of root stocks.

Stenotaphrum secundatum (buffalo grass, crab grass)

Creeping perennial. Spreads by stolons and rhizomes forming a dense turf. Adapted to tropics and sub-tropics. Productive pasture grass for low, moist areas. Useful for erosion control. Rapidly becomes mature and unpalatable. Established from stem cuttings.

Themeda triandra (red oat grass)

Tufted perennial. Natural grassland of East, Central, and South Africa. Maintained by annual burning. Eliminated by overgrazing. Seed rate 22·4–28 kg/ha.

Tripsacum laxum (Guatemala grass)

Tall, leafy perennial which forms large clumps or stools easily uprooted by grazing. Makes good silage or soilage. Drought-resistant. Less productive than elephant grass and lower nutritive value.

LEGUMES

Alysicarpus vaginalis (Alyce clover)

Spreading perennial up to 1 m high. Does not tolerate wet soils. Susceptible to root nematodes. Harvesting difficult as seeds shatter. Seed rate 11–22 kg/ha broadcast. Grown in Ceylon in association with *Brachiaria distachya* and *Paspalum dilatatum*.

A. vaginalis var. *nummularifolius* is used in the West Indies and in Queensland as a pasture legume.

Alysicarpus glumaceus

The most important grazing legume in the low-tree/high-grass savanna of Kenya.

Alysicarpus rugosus

A pasture legume in North America, the Sudan, Queensland, and India.

Calopogonium mucuniodes

Trailing perennial adapted to moist soils. Seed rate 6–9 kg/ha. Forms a dense mat of foliage and re-seeds freely. Grown in association with *Chloris* and *Melinis* grasses. Dies out in dry season. Seed rate in Queensland 3–4·5 kg/ha.

Centrosema pubescens

Prostrate, trailing, leafy perennial. Easily established from seed. Seed rate 4·5–6 kg/ha. Regenerates freely. Grown in association with *Cynodon*, *Chloris*, *Melinis*, *Panicum*, *Andropogon*, and *Pennisetum* spp. Smothers weeds. Crude protein 16–19 per cent of dry matter.

Clitoria ternatea

Prostrate, trailing perennial. Slow to establish and will not tolerate high humidity. Forms a good pasture in association with Rhodes grass (*Chloris gayana*) in the rainlands of central Sudan. Promising pasture legume in Kenya and the West Indies. Seed rate 22–34 kg per hectare.

Desmodium spp.

Perennial semi-woody bushes and vines adapted to moist soils and used as pasture legumes throughout the topics. *D. salicifolium* is grown in association with *Panicum* and *Paspalum* species in tropical Australia. *D. uncinatum* is grown in association with Kikuyu grass (*Pennisetum clandestinum*). *D. adscendens* is important at higher altitudes and in wet areas. *D. barbatum* has shown promise in pastures in Florida and *D. discolor* and *D. canum* in Hawaii and in Queensland. *D. intortum* proved to be the best legume for non-irrigated pastures at Tuchila in Malawi but, being very palatable, tended to be overgrazed. Several other varieties have shown promise in tropical pastures in Malaya, East and West Africa, Central America, Ceylon, India, and Australia.

D. intortum (greenleaf desmodium) and *D. uncinatum* (silverleaf desmodium) are the varieties most commonly grown in Queensland. *D. intortum* is useful in pure stands, in pasture mixtures, and for hay; suffers slightly from little leaf and from *Amnemus* weevil. Grows on most soil types and tolerates poor drainage and flooding. Seed should be inoculated and lime-pelleted. Seed rate 1–3 kg/ha. Should be grazed lightly in first season. *D. uncinatum* similar to *D. intortum* but growing season one month shorter. Seed rate 3–5 kg/ha.

Glycine wightii

Trailing perennial. Vigorous, persistent, drought-resistant, and withstands frost. Seeds well. Mixes well with Rhodes grass in Kenya and with elephant grass in Rhodesia. Also grown in association with Guinea grass, molasses grass, and *Setaria sphacelata.* Is an outstanding pasture legume in Queensland. Seed rate (broadcast) 7–9 kg/ha; (drilled) 4·5–7 kg/ha.

Indigofera spicata

Has shown promise as a pasture legume in tropical Australia Toxicity has been observed due to the presence of the hepatotoxic amino-acid indospicine (Hegarty and Pound 1968.)

Leucaena leucocephala

An evergreen shrub or tree grown as a fodder plant and soil improver. Is often planted in pastures, e.g. interplanted with *Panicum maximum* in Hawaii. Has potential value as a high-protein browse legume. Seeds profusely and is difficult to eradicate.

Lotononis bainesii

Prostrate, stoloniferous perennial. Prefers moist soils. Combines well with several pasture grasses including *Paspalum*, *Digitaria decumbens*, and *Chloris gayana.* Established by seed or cuttings; seed rate 1 kg/ha. Early flowering; frost-tolerant. Promising in subtropical Australia. By far the most palatable of the tropical legumes.

For best results should be kept closely grazed. Does best on light, sandy soils, rainfall 90–100 cm, but can grow on less. Suffers from little leaf.

Medicago sativa (lucerne, alfalfa)

A deep-rooted perennial herb; demands sunshine and adequate lime status; tolerates high temperatures but not high humidity. Grown extensively under irrigation. Sometimes grown in association with pasture grasses, e.g. Rhodes grass in Kenya and Queensland, but is normally grown alone for green fodder and silage. Does not withstand close grazing. Seed rate 11–34 kg/ha broadcast.

Phaseolus lathyroides (phasey bean)

An erect, slightly branching annual-perennial, 120–150 cm high. Combines well with *Paspalum scrobiculatum* in coastal Queensland. Palatable and nutritious; susceptible to frost. Seed rate 2–7 kg/ha.

Phaseolus atropurpureus

In Brisbane in Northern Australia, M. Hutton has succeeded in producing a new cultivar of tropical legume by crossing between two ecotypes of *Phaseolus atropurpureus*. The new cultivar named Siratro is the most adaptable and widely used of the tropical legumes in Queensland. It is a twining vine spreading by runners that root down in the soil. Siratro is a vigorous grower and high producer but has a shorter growing season than *Glycine* and *Desmodium* spp. Establishes readily, competes with weeds, tolerates drought and low-fertility soils; susceptible to bean fly. Seed rate 3–6kg/h.

Pueraria phaseoloides (tropical kudzu)

Prostrate, trailing perennial; stoloniferous. Forms a good pasture in association with molasses grass (*Melinis minutiflora*), Guinea grass (*Panicum maximum*), Dallis grass (*Paspalum dilatatum*), crab grass (*Stenotaphrum secundatum*), Jaragua grass (*Hyperrhenia rufa*), and elephant grass (*Pennisetum purpureum*). Slow to establish. Seed rate 3–4·5 kg/ha. Adapted to high-rainfall regions with a not too prolonged dry season. Does not withstand close grazing.

Stylosanthes guyanensis (Stylo)

If consumed before flowering, this has a nutritive value comparable to the best lucerne. Drought-resistant perennial, persistent, but inclined to be woody. Seed rate in Queensland 2–4·5 kg/ha. *S. guyanensis* is grown in Queensland in association with molasses grass Guinea grass, Para grass, and spear grass (*Heteropogon contortus*). It is a popular pasture legume in Hawaii, Kenya, and Uganda. In Brazil it is grown wherever lucerne will not grow. *S. guyanensis* is ideally suited to poor soils in high-rainfall frost-free areas. Inoculation desirable but not essential.

Stylosanthes humilis (Townsville Stylo)

The only pasture legume of importance in the tropical areas of Australia and now widely distributed in coastal Queensland and in the Northern Territory of Australia. Since its palatability is low it spreads at the expense of the grasses when heavily grazed but when undergrazed is suppressed by grasses. *S. humilis* is an annual or bi-enniel legume easily established without cultivation and adapted to a wide range of soils; prefers 60–90 cm rainfall. Seed rate 2–6 kg/ha.

Tephrosia spp.

Woody, erect, or semi-prostrate shrubs with coarse and unpalatable forage which is often toxic; propagated by seed; used almost exclusively as green manure, cover crop, and soil improver.

Trifolium spp. (clovers)

Annual or perennial herbs, mostly long-day plants found in the humid temperate zones of the northern hemisphere, but a few species are indigenous to South America and Africa. In tropical regions the clovers are grown only at high altitudes, e.g. white clover (*T. repens*) in association with *Paspalum dilatatum* and *Pennisetum clandestinum* in the coastal uplands of Queensland and at 1800–2000m in Uganda where clovers are grown in association with *P. clandestinum*, *Chloris gayana*, *Bromus ciliatus*, and *Phalaris tuberosa*; and under irrigation, e.g. berseem (*T. alexandrinum*) grown extensively in Egypt and the Sudan under irrigation as a soilage crop. *T. cheranganiensis*, *T. johnstonii*, and *T. semipilosum* are important in East African pastures.

Vigna spp. (cowpeas)

Vigorous annual with abundant foliage on trailing or climbing vines. Propagated by seed. Sensitive to cold but resistant to drought and heat; susceptible to fungal attack in humid climates. Certain species, e.g. *V. repens*, have possibilities as pasture legumes, others are grown for food, hay, silage and green manure.

Many tropical legumes yield a high percentage of hard seed so that scarification of the seed by hot water or sulphuric acid is advantageous. Inoculum should be used in the establishment of legumes.

5 Grassland Ecology: Some Tropical Grassland Associations

ECOLOGY may be defined as the responses of a plant population to its environment. The following explanations of the terminology employed in ecological studies are after Sir Bruce Levy (1955). For a more comprehensive treatment of grassland ecology see Appendix A.

Where a *plant community*, comprising any number of plants of the same growth form, becomes widespread, it is called a *plant formation*, e.g. rain forest, tussock grassland. Within the plant formation are smaller groups called *associations*, e.g. the rye-grass/white clover association of temperate regions. These arise from special environmental conditions in certain areas. *Succession* involves a rearrangement of plant associations or of their component species until finally a type of vegetation is evolved that is in harmony with its environment. This is called a *climax* association or climax formation, e.g. rain forest in the humid tropics or desert steppe in the semi-arid regions.

The nature of the plant association is influenced or determined by the following factors: soil fertility, soil moisture, temperature, light and shade, the growth form and seasonal growth of the plant, palatability of the plant to livestock, the incidence of pests and diseases, and weed competition.

The more desirable, high-producing herbage species generally require a high level of soil fertility and will dominate the sward on fertile soils. On soils of low fertility, or when the fertility falls because of faulty management, poor-producing species gain ascendency. Drainage of wet or waterlogged soils and irrigation of dry soils may lead to the replacement of low-yielding species by more productive ones.

Temperature and day length limit the range of many herbage species.

A balance of light and shade is required. The more productive grasses are shade-demanders and are suppressed or eliminated from a pasture by close grazing, which allows light to penetrate to the crown. Resting or rotational grazing with rest periods permits foliage growth and shades the crown. Excessive shade, on the other hand, inhibits growth and in a grass/legume association the legume is readily suppressed by the shading effect of taller grasses.

The growth form determines the persistence of the species. Herbage species may be tall or dwarf, erect or prostrate, stoloniferous or rhizomatous. They may have many or few tillers.

The seasonal spread of a pasture may be extended by the selection of suitable combinations of early-maturing and late-maturing species, by the judicious application of fertilizer, and by manipulation of grazing management.

In addition to edaphic factors the nature of the association may be influenced by biotic factors or human agency. The sward may be improved by the application of deficient nutrients in the form of fertilizer. It may be re-seeded with improved species or improved by controlled grazing and adequate but not too heavy stocking. Fire has a profound influence on grassland associations and, when carefully controlled, is by no means purely destructive. Fire provides the only practicable means of eliminating overmature and useless pastures on many range lands and leads to the production of nutritious regrowth in the rains.

The natural grasslands of the world

The natural grasslands of the world are generally classified into five types: savanna grasslands, grass steppe, shrub steppe, desert scrub, and alpine grasslands according to the nature of the cover (Davies 1959).

Savanna grasslands have more or less complete grass cover with scattered trees sometimes close enough to form woodland but not close enough to shade and so inhibit the growth of the grass. Most tropical grazing lands fall in this category.

Grass steppe consists almost entirely of grass with no legumes and very few herbs. Trees and bushes are usually absent. Natural grasses are of the 'bunch' type. An example is the Argentine pampas.

Shrub steppe is covered with low-growing shrubs often not more than 90 cm tall, e.g. in Patagonia. Animals browse on the shoots and leaves of the shrubs.

In *Desert scrub*, the grass cover consists of annuals which appear after heavy rain. During the dry season animals depend largely on browsing. Desert scrub is typical of the arid and semi-arid tropics.

Alpine grasslands closely related to grass steppe are found in Scotland, the Alps, the Himalayas, etc. The cover consists of short grasses. There are no legumes.

Tropical grasslands

Whyte and Rattray (1958) record sixteen main grass associations in Africa and 100 sub-associations. The main associations are named after the following genera and contain the number of sub-associations noted.

Andropogon	7	*Exotheca*	1
Aristida	8	*Heteropogon*	1
Cenchrus	2	*Hyparrhenia*	30
Chasmopodium	2	*Loudetia*	4
Chloris	1	*Panicum*	3
Chrysopogon	6	*Pennisetum*	3
Ehrharta	1	*Setaria*	2
Eragrostis	7	*Themeda*	22

Whyte has drawn up a table of grass species adapted for cultivation to the main climatic regions of the world. The sub-tropical and tropical sections are quoted in Table 2.

In a footnote Whyte stresses 'that this table is only provisional, that not all the geographical areas may as yet have been included, and particularly, that the inclusion of a species in the third column does not necessarily mean that this particular species is grown or occurs

TABLE 2

Grass species adapted for cultivation to the main climatic regions of the tropics and sub-tropics

Climatic classification	Geographical area	Characteristic natural or cultivated species
TROPICAL		
Extended rainfall	Equatorial belt; islands and littorals north and south of equator	*Brachiaria mutica* *Chloris gayana* *Cynodon dactylon* *Hyparrhenia rufa* *Melinis minutiflora* *Panicum maximum* *Pennisetum clandestinum* (1200–2100 m) *P. purpureum*
Restricted rainfall	Monsoon regions of the tropics; summer-rainfall regions of Central and Southern Africa	*Cenchrus ciliaris* and all the grasses of the preceding column in favourable sites Annual *Sorghum* species
Semi-arid	Monsoon regions with a more restricted rainfall (under 60 cm)	*Cenchrus ciliaris* Annual *Sorghum* species
WARM TEMPERATE OR SUB-TROPICAL		
Extended rainfall	American Gulf region; Queensland and New South Wales littoral; southern China (part); north-eastern Argentina and Uruguay	*Chloris gayana* *Cynodon dactylon* *Paspalum dilatatum* *P. notatum* *Sorghum halepense*
Summer rainfall	South-eastern Great Plains of North America; South African plateau; central and western Argentina; southern China (part); Australia (sub-coastal Queensland, northern coast of New South Wales)	*Andropogon* (*Bothriochloa*) *Ischaemum* *Bouteloua gracilis* *Buchloe dactyloides* *Cenchrus ciliaris* *Chloris gayana* *Eragrostis curvula* *Panicum maximum* *Paspalum dilatatum* *Pennisetum clandestinum* *Sorghum almum* Annual *Sorghum* species

TABLE 2 (*continued*)

Climatic classification	Geographical area	Characteristic natural or cultivated species
Semi-arid	South-western Great Plains of North America; South Africa (part); western Argentina (part); areas south and east of Mediterranean	*Buchloe dactyloides* *Bouteloua curtipendula* *B. gracilis* *Cenchrus ciliaris* *Eragrostis curvula* *E. trichoides*

naturally in all the areas or countries listed in the second column'.

In *The grass cover of Africa*, Rattray (1960) has classified the grass-cover types of African territories as follows:

Congo
- 7 *Hyparrhenia* types
- 3 *Loudetia* types
- 1 *Pennisetum* type
- 2 *Themeda* types

Ghana
- *Andropogon*
- 3 *Hyparrhenia*
- *Pennisetum*

Nigeria
- *Andropogon*
- *Cenchrus*
- 2 *Hyparrhenia*
- *Pennisetum*

Sudan
- 2 *Aristida*
- 2 *Andropogon*
- *Cenchrus*
- *Chloris*
- *Exotheca*
- 2 *Hyparrhenia*
- *Sorghum*

Tanzania
- *Chrysopogon*
- *Eragrostis*
- 2 *Hyparrhenia*
- *Pennisetum*
- 2 *Panicum*
- 3 *Themeda*

Zambia
- *Andropogon*
- 5 *Hyparrhenia*
- 2 *Loudetia*

Rhodesia
- 2 *Andropogon*
- *Cenchrus*
- *Eragrostis*
- 3 *Hyparrhenia*
- *Themeda*

Malawi
- *Exotheca*
- 4 *Hyparrhenia*
- *Themeda*

Uganda
- *Chrysopogon*
- *Cenchrus*
- *Chloris*
- *Exotheca*
- 2 *Hyparrhenia*
- *Loudetia*
- *Pennisetum*
- 2 *Themeda*
- *Setaria*

Forty-three African territories have been classified.

The grassland associations in the tropical African territories of Nigeria and Ghana (West Africa), the Congo, Zambia, Rhodesia, and Malawi (Central Africa), Uganda and Tanzania (East Africa), and the Sudan (North Africa) will be briefly described.

Nigeria (Rains 1956; Keay 1959)

The natural grasslands of Nigeria, which cover more than 80 per cent of the total area of the country, have been classified in the following ecological zones proceeding from north to south.

(1) *Sudan zone*, or *thorn-scrub savanna*, with a rainfall of 64 to 90 cm between May and September. This zone includes the provinces of Sokoto, Katsina, Kano, and Bornu. The grasses are mainly quick-growing annuals which survive the dry season in the seed stage. Browse trees contribute to the fodder. The grass species include *Cenchrus* spp., *Ctenium elegans*, *Schoenefeldia gracilis*, *Eragrostis tremula*, *Aristida* and *Loudetia* spp., *Pennisetum pedicellatum*, *Andropogon gayanus*, and *Andropogon pseudapricus*. The northern part of Bornu Province forms a more extreme zone described as Sahel. The dominant grasses are *Aristida stipoides* and *Schoenefeldia gracilis*.

(2) *Northern Guinea zone*, open woodland with a rainfall of 115 cm between April and October. This zone includes the provinces of Zaria, Bauchi, Northern Niger, and parts of Ilorin Province. The grasses are predominantly perennial with some annuals. They include *Hyparrhenia* spp., *Andropogon gayanus*, *Cymbopogon giganteus*, *Imperata cylindrica*, *Rottboellia exaltata*, *Thelepogon elegans*, *Pennisetum pedicellatum* and *P. polystachyon*, *Digitaria gayana*, *Setaria sphacelata* and *S. pallidefusca*.

(3) *Southern Guinea zone*, or *low-tree savanna*, with a rainfall of 115–150 cm between March and November. This zone includes the provinces of Ilorin, Kabba, Niger, and Benue. The semi-natural grasslands of Western Nigeria lie within this zone. The grasses consist of *Pennisetum purpureum*, *Andropogon tectorum*, and *Panicum maximum* in addition to those of the northern Guinea zone.

Intermediate between the rain-forest and the savanna zones is the *derived savanna zone*, open parkland derived from the forest and maintained by the regular use of fire. The grass cover is mainly tall grasses, predominantly *Andropogon* spp., and herbaceous plants.

The grasses that become dominant in the rain-forest region following prolonged cultivation include *Andropogon* spp., *Pennisetum purpureum*, *Loudetia arundinacea*, *Ctenium newtonii*, and *Imperata cylindrica* (spear grass). There is little bush grazing in the forest region, but productive improved pastures are readily established.

Ghana (Blair 1960)

There are 148 000 square kilometers of grassland in Ghana. The grasslands consist of the interior and coastal savanna zones separated by the rain-forest zone. The interior savanna is made up of the Guinea savanna of the Northern Region and the derived savanna of north Ashanti lying between the Guinea savanna and the rain forest. The dominant grasses in the Guinea savanna are *Andropogon* and *Hyparrhenia* spp. When the grazing intensity is high, these are replaced by shorter grasses consisting largely of *Andropogon pseudapricus*, *Diectomis fastigiata*, *Loudetia togoensis*, *Pennisetum pedicellatum*, and *Heteropogon contortus*. *Aristida kerstingii*, *A. hordacea*, and *Schoenefeldia gracilis* become dominant in areas subjected to extreme grazing pressures.

In the derived savanna, tree cover tends to be denser and, in situations where it approaches a closed canopy, *Andropogon tectorum* is usually dominant, associated with *Beckeropsis uniseta*. In the more open situations and after fire or cultivation *Andropogon gayanus* and *Hyparrhenia subplumosa* may become dominant, with *Pennisetum purpureum* dominant in frequently disturbed areas.

The coast savanna consists of the Accra plains and the Ho-Keta plains lying west and east respectively of the Volta river. Lack of permanent water supplies has left the greater part of the plains undisturbed by grazing or cultivation. On the Accra plains the dominant grasses are short to medium species of *Vetiveria*, *Brachiaria*, *Andropogon*, *Schizachyrium*, and *Monocymbium*. After cultivation on the eastern portion of both plains, taller species of *Andropogon*, *Panicum*, *Hyparrhenia*, and *Pennisetum* become dominant. Tall species of *Andropogon*, *Hyparrhenia*, and *Beckeropsis* dominate the 115–125 cm rainfall belt in the northern portion of the Ho-Keta plains, while medium or short species of *Andropogon*, *Schizachyrium*, *Monocymbium*, *Brachiaria*, *Heteropogon*, and *Sporobolus* are dominant in the drier southern portion.

Congo

The savanna grazing lands of the Nioka district of Congo have been described by Froment (1959). These *Hyparrhenia* savannas are part of the high-altitude savanna of the Lake Albert sector in the eastern Sudano-Zambesian region. The altitude is 1800–2100 m with an annual rainfall of 125 cm.

The *Hyparrhenia* savanna area has ben divided into zones by Taton (1949) and Liben.

(1) The association *Hyparrhenia cymbaria* and *Echinops amplexicaulis*. This association is maintained by bush fires. In their absence it evolves towards a group dominated by *Setaria sphacelata* and eventually reverts to *Albizia* forest. It provides excellent grazing under rational management but, if uncontrolled, is rapidly invaded by *Cymbopogon afronardus*, *Digitaria vestita*, and *Sporobolus pyramidalis*.

(2) The association *Cymbopogon afronardus* and *Eragrostis seticaulis* occurs on the edge of depressions and on more artificial soils than (1). Palatability and carrying capacity are low.

(3) The *Setaria sphacelata* association. This provides good grazing when controlled but otherwise reverts as in (1).

(4) The *Digitaria vestita* association. When overgrazed this association becomes invaded by *Cymbopogon afronardus* and *Sporobolus pyramidalis*. When rationally grazed it becomes richer in *Setaria* and *Hyparrhenia*.

Rotational grazing of the *Hyparrhenia* savanna over ten years at the Institute National pour l'Étude Agronomique du Congo (I.N.E.A.C.) station at Nioka has increased the productivity of the pasture from 22 kg to 157 kg live-weight increase per hectare per year. Each grazing plot is allowed a rest period of two months during the rains and three months during the dry season.

Rhodesia

The grasslands of Matabeleland and Mashonaland have been described by Henkel (1931).

Matabeleland

Two zones are recognized.

(1) *Sweet mixed veld or thorn veld.* The altitude is about 1350 m and the annual rainfall 50–65 cm. There is a long dry season of 8–9 months. The soils are fertile.

The vegetation is largely secondary as this zone has carried a large native population. In the 'undisturbed veld' the important grasses are *Heteropogon contortus*, *Hyparrhenia* spp., *Cymbopogon* spp., *Themeda triandra*, *Setaria* spp., *Amphilophis insculpta*, *Brachiaria* spp. and *Digitaria* spp. The sweet mixed veld provides good winter grazing.

(2) *Sour mixed veld or Mangwe veld.* The soils of the sour mixed veld are infertile, sandy, and acid. Characteristic herbage plants in the 'undisturbed veld' are:

Hyparrhenia spp.
Cymbopogon spp.
Heteropogon contortus
Andropogon spp.
Elyonurus spp.
Digitaria spp.
Pogonarthria squarrossa
Schizachyrium jeffreysii
Crossotropis grandiglumis
Trichopteryx (*Loudetia*) *simplex*
Perotis indica.

Winter grazing is very poor and cattle are wintered on the adjoining thorn veld when possible.

Mashonaland

The soil is sandy with patches of red, fertile clay. The grasslands have been classified into four zones.

(1) *Eastern mountain range.* The altitude ranges from 1800 to 2150 m; the annual rainfall is 100–180 m. This zone consists of mountain grassland characterized by a good cover of short grasses. *Themeda*

triandra, *Hyparrhenia* spp., *Eragrostis* spp., and *Trichopteryx* (*Loudetia*) are the dominant grasses.

(2) *High veld.* Altitude 1400–1800 m; annual rainfall 75–100 cm. The valley grassland is known as vlei grassland.

(3) *Middle veld.* Altitude 700–1400 m. The grasses are tall and coarse but are generally of better quality than those of the more heavily leached high-veld soils.

(4) *Low veld.* Altitude below 700 m with rainfall less than 50 cm. On sandy soils the grass tends to be open and the plants tufted, while the alluvial soils frequently have a good cover of sweet grasses giving valuable winter grazing.

The sweet veld occurs in the drier and more fertile areas at low altitudes and generally contains grasses that are palatable and nutritious in winter as well as summer seasons. The sour veld occurs at much higher altitude with a higher rainfall. The grasses mature readily and are palatable only in the young stage. Good natural grazing is thus confined to the growing season. Mixed veld is intermediate between sweet and sour veld and provides grazing throughout the year.

Zambia (Rattray 1960)

A sparse, predominantly annual grass cover composed of *Aristida* and *Eragrostis* species together with other annuals and a few perennials such as *Andropogon gayanus* and *Hyparrhenia* species is associated with woodland or tree savanna at 350–1000 m.

There are five dominant *Hyparrhenia* communities at altitudes of from 500–1800 m and on the Kalahari sands at 1000–1800 m above sea level, two *Loudetia–Tristachya* communities, one with *Hyparrhenia dissoluta* and the other with *Monocymbium ceresiiforme* as conspicuous associated species.

Precise information is lacking on grassland communities of the flood-plains. *Themeda triandra* is often conspicuous in seasonal swamp areas, which also contain *Echinochloa* spp. and *Vossia cuspidata*. These swamp areas provide good dry-season grazing.

Trapnell (1950, 1953) has classified the flood-plain vegetation of

Zambia in three categories according to soil type: (1) black mangalitic clays—*Hyparrhenia rufa* and *Setaria* spp; (2) sandy type—*Loudetia* spp. and *Themeda triandra*; (3) alluvial soils—*Vetiveria nigritiana* (on bunds) and *Echinochloa* spp. (on flooded areas).

Malawi (Rattray 1960)

Rattray lists four *Hyparrhenia* dominant grassland communities associated with woodland from altitudes of less than 700 m up to 1800 m. Associated grasses, in various combinations, are *Andropogon* spp., *Setaria* spp., *Panicum maximum*, *Urochloa* spp., *Themeda triandra*, *Digitaria* spp., *Schmidtia bulbosa*, *Euclasta condylotricha*, *Eustachys paspaloides*, *Thyrsia undulatifolia*, *Bewsia biflora*, and *Eragrostis superba*.

Between 1800 and 2100 m there is a grassland community composed of *Themeda triandra* in association with *Exotheca abyssinica*, *Monocymbium ceresiiforme*, *Elyonurus argenteus*, *Brachiaria serrata*, *Andropogon schirensis*, *Hyparrhenia lecomtei*, and *Loudetia simplex*.

A short, tufted montane grass land consisting of *Exotheca abyssinica*, *Loudetia simplex*, *Trachypogon spicatus*, *Helictotrichon elongatum*. *Agrostis* spp., *Monocymbium ceresiiforme*, and *Elyonurus argenteus* lies between 2180 and 2800 m above sea level. *Festuca schimperiana* is common in the wetter areas and *Danthonia davyi* in exposed situations. The nutritional value of this montane grazing falls off in the cold season.

There are also seasonal swamps with grass/sedge cover characterized by communities such as *Hyparrhenia rufa*/*H. nyassae;* or *Chloris gayana*/*Dichanthium papillosum*/*Cynodon plectostachyus*/*Bothriochloa glabra*; or *Ischaemum afrum*/*Setaria palustris*; or *Panicum repens*/*Sacciolepis* spp./*Leersia hexandra*.

Uganda (Thomas 1940)

The grasslands of Uganda have been divided into three zones.

(1) *Long-grass zone*. This zone includes those parts of Uganda between 1000 and 1800 m with a fairly evenly distributed annual rainfall of 115 cm. The dominant grass is elephant grass (*Pennisetum purpureum*) which, when undisturbed, may grow to 3 m or more. Overgrazing leads to invasion by *Imperata cylindrica* (spear grass) and

'lumbugu' (*Digitaria scalarum*), a pernicious weed which is palatable and nutritious but extremely difficult to eradicate. The spear grass may be eliminated by repeated cutting and, although generally considered undesirable, does provide useful grazing when young.

(2) *Short-grass zone*, This zone includes the rest of the country below 1800 m with an annual rainfall of less than 115 cm. The dominant grass is *Hyparrhenia* with *Brachiaria* spp. as bottom-filling grasses. *Setaria sphacelata* and *Chloris gayana* are widely distributed. *Panicum maximum*, *Sporobolus* spp., and *Andropogon* spp. often achieve local dominance. *Cynodon dactylon* and *Eragrostis* spp. are classed among the common weed grasses of the zone but both of these can provide valuable grazing.

(3) *Kikuyu-grass zone*. This zone occurs at 1800–2100 m on Mount Elgon and in Kigezi. The pastures are predominantly Kikuyu grass (*Pennisetum clandestinum*) mixed with legume *Trifolium johnstonii*. These valuable pastures are able to withstand hard grazing.

In Uganda the natural grasslands are burnt over each year during the dry season in order to eliminate coarse, overmature grasses and to obtain fresh growth for grazing. In Teso the first grass to reappear is spear grass followed by *Setaria sphacelata* and *Pennisetum polystachyon*, *Brachiaria brizantha* and *B. fulna*, *Hyparrhenia* spp., *Urochloa* spp., *Paspalum* spp., *Sporobolus* spp., *Panicum maximum* and *Trichopteryx* (*Loudetia*) spp. in order of appearance. With the first rains *Digitaria* spp., *Chloris* spp., and *Eragrostis* spp. reappear.

Tanzania

Van Rensburg (1956) has classified the major vegetation types in Tanzania as follows.

(1) *Deciduous thicket*, which does not support natural grassland.

(2) *Deciduous woodland* locally referred to as 'miombo'. The rainfall averages about 95 cm per year. The grasses of this zone are predominantly *Hyparrhenia dissoluta*, *H. filipendula*, *Rhynchelytrum repens*. and *Pennisetum polystachyon*. On the more fertile alluvial soils

of this vegetation zone a mixed grassland flora containing *Andropogon gayanus*, *Hyparrhenia rufa*, *Echinochloa pyramidalis*, and *Chloris gayana* occurs. Under more favourable conditions, for example along some of the far slopes of mountain ranges, *Panicum maximum* and *Mucuna pruriens* occur. Under conditions of controlled grazing the tall bunch grasses give way to shorter grasses, e.g. *Cynodon plectostachyus*, *Urochloa pullulans*, *Digitaria milanjiana*, and *Chloris gayana*.

(3) *Mountain grassland*, typical of areas above 1800 m with annual rainfall exceeding 75 cm. The grassland is maintained by periodic burning. In rain–forest areas the dominant grass is *Pennisetum clandestinum*. On the red loam soils of medium fertility *Themeda triandra* is generally dominant. On granitic and sandstone soils of low fertility *Themeda triandra* gives way to *Loudetia simplex* and other species including *Eragrostis chalcantha*, *Hyparrhenia hirta*, and *Melinis minutiflora*. On shallow lateritic soils *Loudetia kagerensis* is the dominant species.

The Republic of the Sudan (Snow 1948)

The nomadic grazing pastures of the Sudan are located in the Kordofan and Darfur provinces, an area of plains and stabilized sand-dunes covered by many grasses in amounts governed by the rainfall. The annual rainfall varies from 30 to 50 cm. The most common fodder grasses are probably species of *Aristida*, tussock grasses that produce little leaf and seed and are of low nutritive value. Other grasses of the plains are species of *Sporobolus*, *Eragrostis*, *Brachiaria*, *Setaria*, and *Chloris*, *Schoenefeldia gracilis*, and *Dactyloctenium aegyptium*. *Cenchrus* species are found on the sand-dunes and *Echinochloa colonum* and *Cynodon dactylon* on the edges of pools on heavy soil.

The grazing lands of western Equatoria in the southern Sudan (Catford 1952), which have a rainfall of 114–138 cm at an altitude of 700–1000 m, contain *Pennisetum purpureum*, *Hyparrhenia* spp., *Panicum maximum*, and *Imperata cylindrica* as dominant grasses. *Chloris gayana* and *Setaria sphacelata* are rare. In general this region is comparable with the Uganda short-grass zone and the Nigeria southern Guinea zone.

Australia (Christian and Donald 1949)

The pastures of the tropical north of Australia include the following species: *Bothriochloa, Dichanthium, Heteropogon, Chrysopogon, Iseilema, Themeda, Panicum, Brachiaria, Chloris* and *Astrebla.*

Most of the indigenous perennial grasses form small tussocks separated by bare soil; rhizomatous species are rare.

Queensland.

The native pastures of Queensland in tropical northern Australia have been described by White (1934) and McTaggart (1935).

(1) *North, coastal, and central Queensland.* The grasses include species of *Andropogon, Aristida, Arundinella, Astrebla, Brachiaria, Chloris, Cynodon, Eragrostis, Iseilema, Ischaemum, Panicum, Paspalum, Paspalidium, Rottboellia, Setaria, Sporobolus, Stipa,* and *Sorghum,* and *Dichanthium sericeum, Amphilophis intermedia, Cymbopogon procerus, Heteropogon contortus,* and *Themeda australis.* Among legumes are *Stylosanthes sundaica* (*humilis*) and *Alysicarpus* spp.

The freshwater swamp pastures of coastal Queensland are of high grazing value. Species include *Paspalum distichum, Panicum obseptum, Leersia hexandra, Echinochloa crusgalli,* and *Hemarthria compressa.* The salt-water meadows of the coastal belt consist of *Sporobolus virginicus* var. *minor* and, in the more muddy places towards the edge of the pasture, *Paspalum vaginatum.*

(2) *Western Queensland.* These high-quality pastures contain *Astrebla* spp. (Mitchell grass), *Iseilema* spp. (Flinders grass), *Dichanthium sericeum* (blue grass), and species of *Panicum, Eragrostis,* and *Chloris.*

R. Milton Moore (1970) has compiled maps of the grazing lands, the vegetation, and the pastures of Australia. In the same volume (p. 112), Shaw and Norman describe the tropical and sub-tropical grasslands of Australia. The tropical grazing lands are classified as 'tropical tall grass'. Major species of the tropical tall-grass area in the Northern Territory include *Themeda australis, Sorghum plumosum, S. citrans,* and *S. stipoideum, Astrebla* spp., *Eulalia fulva, Iseilema* spp., *Plectrachne pungens*, and *Eriachne burkitii* (Perry 1960). The Western Australia region is dominated by *Plectrachne pungens*, *Chrysopogon fallax*, *Sehima nervosum*, *Sorghum* spp., and *Aristida hygrometrica* (Speck *et al.* 1960). In Queensland's tropical tall-grass zone the dominant species is *Heteropogon contortus* (bunch spear grass). Other prominent species are *Themeda australis*, *Bothriochloa* spp., and *Aristida* spp. Species of the genera *Chrysopogon*, *Cymbopogon*, *Chloris*, *Dichanthium*, *Eriachne*, *Schizachyrium*, *Sorghum*, and *Eragrostis* are common.

The *Dichanthium* (blue-grass) grasslands occur mainly in the drier western half of the woodland region. In association with the dominant *Dichanthium* species are grasses from the genera *Aristida*, *Bothriochloa*, *Enneapogon*, *Panicum* and *Thellungia*. Native legumes are common, the main species being *Rhynchosia*, *Glycine*, *Neptunia*, *Alysicarpus*, and *Crotalaria*. The blue-grass areas provide better grazing than the spear-grass areas.

West Indies (Motta 1956)

The former British West Indian islands consist of Trinidad, Grenada, St Vincent, St Lucia, Dominica, Antigua, St Kitts, Nevis, Montserrat, British Virgin Islands, Barbados, Jamaica, the Bahamas, and a few smaller islands. They lie within the tropics between 10° and 20° North latitude. The average annual rainfall varies between 76 and 500 cm.

Few, if any, of the grasses found in these territories are indigenous. The origin of most species is Africa. The species most commonly found in the pastures of the eastern Caribbean areas are *Axonopus compressus*, *Cynodon dactylon*, *Andropogon nodosus*, *A. caricosus*, several *Paspalum* species, *Panicum maximum*, and *Brachiaria mutica*. The *Andropogon* spp. and *Cynodon dactylon* are mainly found in the drier parts of the islands, while *Axonopus compressus* dominates the high-rainfall areas. *Panicum maximum* is found on the better-drained slopes. *Paspalum* spp. have a wide range of adaptability.

Hawaii (Beddows 1944)

The Hawaiian Islands show great variation in topography, altitude, soils and rainfall; the soils vary from new lava and young soils to well-weathered lateritic types and the rainfall from 25 cm to over 630 cm per annum.

The vegetation has been classified in five zones. The number of important forage species is relatively few.

(1) *Zone A*. Altitude, sea level to 700 m mean temperature 23·9 °C, annual rainfall 30–50 cm with long dry periods.

The indigenous grasses include *Setaria verticillata*, *Chloris inflata*, *C. virgata*, *Panicum* spp., and *Heteropogon contortus*. Improvement is limited by the aridity of this zone.

(2) *Zone B*. Average altitude about 700 m annual rainfall up to 100 cm or more. The grasses include those listed under Zone A and *Digitaria* spp., *Cenchrus echinatus*, *Eragrostis amabilis*, *Rhynchelytrum repens*, and *Trichachne* (*Digitaria*) *insularis*.

(3) *Zone C*. Ranges from sea level to 1400 m. The most valuable grass is *Cynodon dactylon*. On poorer soils are found *Chrysopogon aciculatus*, *Setaria geniculata*, and *Paspalum orbiculare*, and on the wetter areas *Sporobolus capensis* tends to become dominant. *Rhynchelytrum repens* is a volunteer on cultivated land.

(4) *Zone D*. This zone, situated on the windward side of the high mountains, has adequate or excessive rainfall. Cool temperatures and frequent fogs produce humid temperature conditions and the vegetation consists largely of species found in temperate regions.

(5) *Zone E*. Altitude 1400–3600 m on the islands of Maui and Hawaii. The indigenous grasses include *Trisetum glomeratum*, *Panicum tenuifolium*, *Eragrostis leptophylla*, *Deschampsia rubigena*, *Festuca dertonhensis*, and *Sporobolus capensis* (*S. africanus*).

Mexico (Buller 1960)

The Mexican tropics have been divided by Buller into ecological grassland types after Leopold's (1950) study of the vegetation zones of Mexico.

(1) *Tropical deciduous forest*. In this zone, after disturbance, *Bouteloua filiformis* is one of the most widely distributed native grasses on the well-drained soils. *B. hirsuta* and *B. curtipendula* also

occur. Other species found in overgrazed pastures are *Opizia stolonifera*, *Sporobolus capensis*, *Andropogon bicornis*, and *Panicum trichoides*. Associated herbaceous legumes are species of *Sesbania*, *Dolichos*, *Phaseolus*, *Crotalaria*, *Calopogonium*, *Teramnus*, and *Indigofera*.

(2) *Disturbed tropical evergreen forest*. This region has a high and well-distributed rainfall. After clearing, a productive pasture including *Paspalum notatum*, *P. conjugatum*, *P. minus*, *Axonopus affinis*, and *A. compressus is developed.*

(3) *Savanna grasslands*. This represents only about 1 per cent of the tropical grass land area. The common grass species are *Imperata brasiliensis*, *Trichachne* (*Digitaria*) *insularis*, *Paspalum virgatum*, *Andropogon bicornis*, *A. glomeratus*, and *Homolepis aturensis*.

(4) *Grasses of the moist river basins and flooded areas*. The valuable native forage species found in these localities is *Leersia hexandra*, while some of the flooded areas are invaded by *Paspalum fasciculatum* and many species of Cyperaceae and Musaceae.

Latin America (Roseveare 1948).

The grasslands of the hot tropical territories of Latin America consist of more or less arid grazings and the savannas of hot climates. The former are used for cattle and sheep ranching and are found in

(a) north-eastern Brazil;
(b) the western, dry pampa and the monte of Argentina;
(c) central Mexico;
(d) central Brazil.

The hot climate savannas, which appear to be suitable only for native cattle and their crosses because of the excessive heat and the incidence of pests, include

(a) the flood-plains or 'llanos' of Venezuela and Colombia;
(b) the Bolivar savannas of Colombia and the lowland savannas of Costa Rica;
(c) savannas of the Amazon basin, Guyana, and Surinam;
(d) savannas of Bolivia, Mato Grosso, and the Gran Chaco;
(e) the Cuban savannas.

The tropical grasslands of Mexico have been described. Those of north-eastern and central Brazil, the Brazilian state of Mato Grosso, and the llanos of Venezuela and Colombia are outlined below.

North-eastern Brazil

About 35 per cent of the region lies in the zone of equatorial climate and 65 per cent in the zone of sub-tropical climate (all within the area classified as tropical by the author). The average of the temperature maxima is 35·5 °C and the average of the minima is 18·8 °C. There are pronounced wet and dry seasons. The annual rainfall ranges from 25 cm to over 65 cm.

The natural grazings include the following: *Gymnopogon mollis*, *Aristida adscensionis*, and *Digitaria sanguinalis* with Cactaceae and browse plants.

Central Brazil

The natural pastures have been improved by planting *Melinis minutiflora*, *Hyparrhenia rufa*, and *Panicum maximum*. Associated with these grasses are *Chloris gayana* and *Pennisetum clandestinum*.

Mato Grosso

The Brazilian state of Mato Grosso, used extensively for stock-raising, is an example of hot-climate savanna.

The important grasses are species of *Panicum*, *Paspalum*, *Eragrostis*, *Andropogon*, *Chloris*, *Manisuris*, and *Sporobolus*. In the uplands there are wide areas in which *Tristachya* species are dominant. Of legumes *Desmodium* spp. are abundant and species of *Crotalaria* and *Stylosanthes* are also found.

The floodplains or llanos of Venezuela and Colombia

(1) *Venezuela*. The llanos lie to the north of the Orinoco river and slope gradually from the base of the Andes to the river. They seldom exceed an altitude of 215 m above sea level, The mean annual rainfall ranges from 80 to 180 cm and the temperature from 20 °C to 28 °C. Three types of savanna are represented: open savanna (grass alone), orchard savanna (scattered bushes), and palm savanna with occasional fan palms.

Pittier (1942) lists as dominant families in the central llanos not far from Caracas the following grasses: *Paspalum* (nine species), *Andropogon* (nine species), *Panicum* (six species), *Eragrostis* and *Sporobolus* (four species each), and the legume *Desmodium* (eleven species).

In the eastern llanos the dominant grasses found in extensive areas were *Trachypogon plumosus*, *T. montufari*. *Aristida riparia*, *A. cognata*, *Paspalum gardnerianum*, *Axonopus anceps*, and *Arundinella hispida*. These are coarse grasses, probably dominant as a result of annual burning. Mixed with these are the more valuable *Andropogon hirtiflorus*, *A. sylvanus*, *A. bicornis*, and a few other species: *Elyonurus adustus*, *Sporobolus cubensis*, and *Axonopus purpusii*.

Pittier subsequently classified the vegetation of the Mesa de Guanipa, a high tableland in this region, as follows.

Arrow-grass savannas. These comprise forty-four species of grasses of which five, *Leptocoryphium lanatum*, *Trachypogon vestitus*, *T. montufari*, *T. plumosus*, and *Axonopus chrysodactylon* are dominant. The *Trachypogon* species are not eaten by stock because of the penetrating point at the base of the florets, which gives rise to the name.

Periodically flooded savannas form a large plain on the left bank of the river Cáni. The *Trachypogon*, *Axonopus*, and *Aristida* species of the arrow-grass savanna are rare and are replaced by finer grasses including *Paspalum* (five species) and *Mesosetum chaseae*.

Palm savannas are woodland formations that accompany the rivers of the eastern llanos and indicate moist conditions. Six species of *Paspalum* and the legumes *Desmodium adscendens* and *D. barbatum* are listed. There are also 'dry' palm groves where *Aristida setifolia* is abundant in the open spaces (Pittier 1942).

(2) *Colombia*. The Colombian part of the llanos lies east of the Andean range known as the Eastern Cordillera. The climate and other environmental conditions of the Colombian llanos closely resemble those of Venezuelan llanos, and the plant cover is similar. Like the Venezuelan llanos, these natural grasslands are composed of indigenous grasses of low nutritive value which soon became fibrous.

They are burned periodically, the fresh shoots being re-grazed. The useful grasses include *Panicum purpurascens* (Para grass), *Panicum maximum*, and *Hyparrhenia rufa*.

The grasslands of Colombia have been reviewed by Crowder (1967).

In addition to the above typically tropical grasslands there are in the tropical region of Latin America vast areas of cool, mountain grasslands whose herbage more closely resembles that of temperate regions. Examples of cool, mountain grasslands are found in the Andean grazings of central Chile, Peru, and Ecuador, the mountain grasslands of Colombia and Venezuela, western Argentina and southern Chile, the upland pastures of the central plateau of Costa Rica, and the paramos. Cold-climate arid grazings comprise the greater part of Patagonia.

The Andean grazings of Peru and Ecuador

Peru (Hitchcock 1927)

The native grasses in the uplands of Peru produce abundant forage for sheep and cattle. The dominant grass over large areas is the bunch grass ichu (*Stipa ichu*), which is grazed by stock only when the leaves are young. Sheep graze *Muehlenbergia* (*Sporobolus*) *ligularis* which grows between the bunches of ichu. The high altitude forage species belong chiefly to *Festuca*, *Calamagrostis*, and *Poa*, and some wild clover (*Trifolium amabile*) is found in cultivated areas.

In the south of Peru in the Department of Puno (4650 m) the principal grasses are *Distichlis* spp. and *Sporobolus* spp. with *Bromus catharticus*, *Festuca dissitiflora*, *Poa annua*, and *Paspalum* spp. Temperate-region grasses, e.g. perennial and Italian rye-grass, cocksfoot, the fescues, and wild clovers form good temporary leys on the more fertile soils on sheltered sites.

Ecuador

Stock-raising is concentrated mainly in the temperate inter-Andean tableland, 1400–4600 m above sea level, and in the mountain grazings above this zone.

Deyeuxia (*Calamagrostis*) species together with species of *Festuca* form immense swards covering the upper part of the Cordilleras.

These provide good grazing. Some forms of *Agrostis*, *Polypogon*, *Holcus lanatus*, *Corynephorus canescens*, *Phleum pratense*, *Dactylis glomerata*, *Lolium* spp., *Anthoxanthum odoratum*, and *Sporobolus tenacissimus* and the legumes lucerne (*Medicago sativa*) and some varieties of clover are cultivated in the mountains. At very high altitudes *Agrostis* species have been recorded and at intermediate altitudes there are grass communities with *Eragrostis densissima* dominant together with *Aegropogon cenchroides*, an *Agrostis*, and a *Paspalum* species. *Andropogon argenteus* was found at about 4300 m.

India

The types of grassland cover of India were described briefly by Dabadghao at the 8th International Grassland Congress, 1960. Five types of grassland cover are recognized.

(1) *Dichanthium/Cenchrus/ Elyonurus*. This type of cover lies between 23° and 32° N and 60° and 80° E. The annual rainfall varies from 10 to 76 cm. Summer temperatures are high while winter temperatures sometimes fall below freezing point. The soils are alluvial, sandy to sandy loam in texture.

The principal perennial grasses are, in order of importance, *Dichanthium annulatum*, *Cenchrus ciliaris*, *C. setigerus*, *Elyonurus hirsutus*, *Eleusine flagellifera*, *Cynodon dactylon*, *Sporobolus pallidus*, and *Panicum antidotale*. The annual grasses include species of *Aristida*, *Echinochloa*, *Eragrostis*, *Brachiaria*, and *Tragus*.

(2) *Sehima/Dichanthium*. This cover is found throughout the tropical portion of India below the tropic of Cancer. Annual rainfall ranges from 38 cm to over 200 cm. All shades of black and red soil are found.

The principal perennial grasses, in order of importance, are *Sehima nervosum*, *Chrysopogon montanus*, *Themeda triandra*, *Eremopogon foveolatus*, *Heteropogon contortus*, and *Cymbopogon* spp. Annual grasses are represented by *Themeda quadrivalvis*, *Apluda aristata*, *Capillipedium* spp., *Pseudanthistiria hispida*, *P. heteroclita*, *Diectomis fastigiata*, *Schizachyrium exile*, and *S. brevifolium*. On deep soils in the valleys or on plateaux the grasses in order of importance are *Dichanthium annulatum*, *Bothriochloa pertusa*, *Eremopogon foveolatus*, and *Cynodon dactylon*. In moist situations, *Dichanthium caricosum*, *Iseilema laxum*, *I. wightii*, and *Ischaemum aristatum* are found. Associated pasture legumes include *Atylosia scarabaeoides*, *Desmodium diffusum*, *Alysicarpus rugosus*, *A. longifolius*, *Heylandia latabrosa*, and *Indigofera glandulosa*.

(3) *Phragmites/Saccharum*. This type occurs in the eastern humid portion of the great lowland plains of the Ganges and Brahmaputra valleys and extends into the first cover described above. The principal grass species are *Phragmites karka*, *Saccharum* spp., *Imperata cylindrica*, *Desmostachya bipinnata*, *Vetiveria zizanioides*, and *Narenga porphyrocoma*.

(4) *Themeda/Arundinella*. Occurs in the northern mountain belt from 530 m to over 2800 m. The principal grass species are *Themeda anathera*, *Arundinella* spp., *Eulaliopsis binata*, *Chrysopogon* spp., *Dimeria* spp., *Bothriochloa* spp., *Heteropogon contortus*, and *Pennisetum orientale*. *Microstegium ciliatum* is an important annual. Legumes are represented by the genera *Desmodium*, *Crotalaria*, *Lespedeza*, *Pueraria*, *etc*.

(5) *Temperate and alpine cover*. This cover contains temperate genera.

Types (1) and (2) are known for cattle and sheep-raising but are badly deteriorated owing to overgrazing. Type (3) is grazed only for the early spring growth after burning. Type (4) is mainly cut for hay.

6 Types of Grazing Pasture

GRAZING pasture may be classified into four main groups.

Natural or semi-natural grasslands

Consequent herbage species in this group have not been sown or planted and the flora has been relatively undisturbed by human agency. The only human interference is the control of grazing animals, generally by herding, and annual or less frequent burning. Most tropical grazing lands come within this category.

The natural grasslands are, in general, climax or sub-climax types that have evolved in harmony with soil, climate, and environmental factors including fire and under a system of lenient grazing or spasmodic heavy grazing by migratory herds of livestock. It is not unlikely that the natural grasslands of the tropics, when improved and adequately exploited, will form major centres of livestock production in the world. At present, these grasslands carry nearly half of the grazing livestock and produce one-third of the meat and one-sixth of the dairy products of the world (Davies 1960). Methods available for the improvement of these grasslands are described in Chapter 12. Of primary importance are the introduction of suitable legumes to form grass/legume associations, and the control of grazing.

Improved permanent grasslands

Constituent herbage species in this group have not been sown or planted, but the botanical composition has been modified in favour of more productive species by careful control of grazing or by mowing, drainage, the application of fertilizers, surface cultivation, re-seeding, and weed control. Other factors that influence the botanical composition of the sward are rainfall and altitude and the exposure of the site. The type of grazing stock also has an effect.

In the United Kingdom, New Zealand, Holland, and other intensively farmed countries of temperate regions, grassland farming has progressed to a degree where it is possible to classify improved permanent pastures according to their botanical composition. Thus in the United Kingdom these pastures are graded according to the proportion in the sward of three species, rye-grass (*Lolium perenne*), bent (*Agrostis* spp.), and wild white clover (*Trifolium* spp.).

First-grade pasture : rye grass/clover;
Second-grade pasture : rye grass/*Agrostis*/clover;
Third-grade pasture *Agrostis*/rye grass/clover;
Fourth-grade pasture *Agrostis*/clover;
Fifth-grade pasture *Agrostis*.

The best-quality fattening pastures of the United Kingdom belong to the first grade, while the *Agrostis* sward is found on poor, infertile soils whose grass cover has been mismanaged by overgrazing in winter and/or undergrazing in summer. In association with the major species listed there are minor proportions of other species.

Tropical grassland husbandry is far from this stage of development, although several productive grass and grass/legume associations have been evolved in different parts of the tropics, e.g. giant star grass/*Centrosema* in Nigeria and molasses grass/tropical kudzu in Puerto Rico.

Artificial grasslands or leys (temporary pastures)

The herbage plants have been sown or planted. Long-duration leys are generally described as permanent pastures. Short-duration leys which may be down for three or four years or less are frequently included in crop rotations for the purpose of restoring soil fertility, e.g. the three-year elephant-grass (*Pennisetum purpureum*) leys, which, when alternated with three or four years of cropping, have led to considerable soil improvement in Uganda.

Irrigated pastures

Naturally irrigated pastures occur along the White Nile and its tributaries in the Sudan and along the Zambesi and its tributaries in Rhodesia. These provide dry-season grazing for migratory herds.

Irrigation of pastures and fodder crops is a normal practice of arid lands where a water source is available and, with the introduction of the sprinkler method, is now being extended to other areas where rainfall is inadequate for part of the growing season. The method adopted depends on the source of the water supply and the nature of the terrain, and may be 'surface', or 'flooding'.

The cost of large-scale irrigation is beyond the means of most graziers in the tropics, where the value of animal produce is low. In these circumstances artificially irrigated land is most advantageously used for the provision of soilage legumes and other fodder crops to supplement the poor-quality dry-season grazing. At the University of Khartoum farm, berseem (*Trifolium alexandrinum*) grown under irrigation and fed as soilage forms the major part of the ration for a dairy herd of Zebu (Butana and Kenana) cattle with an average milk production approaching 2270 litres per head per lactation.

In the tropics it is possible with irrigation to maintain green pasture and fodder crops throughout the year, but the low return from animal produce does not justify the expenditure except in the case of intensively managed urban dairies or fattening units. In Tanzania, at Mpwapwa, irrigated elephant grass (*Pennisetum purpureum*) has yielded over 250 tons of green fodder per hectare per year with a crude protein content of 10 per cent of the dry matter, and lucerne has yielded 88–112 tons green material per hectare with 20 per cent crude protein in the dry matter.

When, under intensive management, irrigated pastures are used for grazing rather than for soilage, or feeding of cut herbage, the system adopted must be rotational, preferably with a movable electric fence, and the pasture should not be grazed until a least 2–4 days after watering when the surface has dried out sufficiently to prevent damage from trampling. Rotational or strip grazing also gives a measure of control of bloat. In order to maintain high production on irrigated pasture, high rates of fertilization are necessary.

7 Seed Production

FEW tropical grasses seed freely, and when they do the seed is often of low viability so that vegetative propagation by stem cuttings or root division is generally necessary. Selection and breeding of free-seeding species is, however, being undertaken, especially in Queensland, Kenya, and Rhodesia. It has been reported that the percentage of seed-producing plants in *Brachiaria brizantha* was increased from 3 to 25 per cent of the total population by repeated selection and re-planting, in successive years, of those plants that did produce seed (Whyte, Moir, and Cooper 1959).

Improved strains of Rhodes grass (*Chloris gayana*), now grown for seed in Kenya, produce 225–550 kg of seed per hectare. In temperate countries the tendency is to select and breed for more leafy strains of grasses, the leaf being more nutritious than the mature stem and the greater vegetative growth leading to extension of the grazing season. The aim in tropical countries is to evolve better seed-producing species so that the expensive and tedious process of vegetative propagation may be replaced by seed drilling or broadcasting.

Since herbage-grass species are naturally cross-pollinated, it is essential to ensure safety from contamination with foreign pollen by isolation of the field on which the seed-crop is grown. A distance of 200–300 metres from other fields containing herbage species is probably adequate. Precautions against viable seeds left in the soil after other grass crops have been grown must also be taken, and the previous history of the field must be known. Fertility of the soil is important. Herbage seed-crops require high fertility, the grass species responding to nitrogen and legumes to phosphate and potash. The seed-bed should be firm, and rolling before sowing is often desirable. Small grass seeds should be sown in shallow drills and not covered too deeply.

Commercial strains of grasses are generally grown in pure stands under a nurse or companion crop, e.g. Rhodes grass (*Chloris gayana*) is often grown under a maize crop. They may be harvested by a binder and threshed with an ordinary threshing machine. Pedigree strains are normally grown in widely spaced rows, 60 cm apart, without a nurse crop. Seed rates are low; 3–7 kg per hectare. Crops may be cut and bound or combine-harvested. Seed is collected in the second and third year and sometimes also in the fourth year after sowing.

For storage the seed should be dried to 10–12 per cent moisture content and stored at a temperature in the vicinity of the freezing point. Dusting with a mercurial before storage is a wise precaution against mould and weevils.

Seed quality

Poor seed is always a bad investment and, in the long run, may prove much more costly than higher-priced seed of known purity and germination. Several countries excercise legal control over the minimum requirements for seed quality in grasses and legumes. Seed merchants generally give guarantees to purchasers regarding purity and germination.

Factors that affect germination and vigour are age, storage conditions, disease, dormancy, hard seeds, and abnormal seeds.

Impurities that may occur in seeds are inert matter, seeds from other crops, and weed seeds.

Seed certification

Seed certification was pioneered by the New Zealand Department of Agriculture in 1929–30 when it introduced a seed-certification scheme for the strain of perennial rye–grass (*Lolium perenne*) that had evolved naturally in the Hawkes Bay and Poverty Bay districts of the North Island of New Zealand. The benefits of the scheme were soon appreciated by the farming community and it was subsequently extended to include Italian or short-duration rye-grass (*L. multiflorum*), cocksfoot (*Dactylis glomerata*), and clovers. The scheme was designed to identify seed from areas of proved strain and to ensure that the quality of a strain was maintained from generation to generation.

Similar schemes have since been introduced in other countries, and in Rhodesia the Rhodesia Pasture Seed Growers' Association now issues certified seed of the following tropical pasture grasses: *Cenchrus ciliaris*, *Chloris gayana*, *Cynodon plectostachyus*, *Eragrostis curvula*, *Panicum coloratum*, *Pennisetum purpureum*, *Setaria sphacelata*, and *Sorghum verticilliflorum*.

Selection and Breeding

Hutton in concise reviews (1964, 1965, 1968*a*) has outlined the objectives of the breeder of tropical pasture plants. In the case of tropical grasses yield is of less importance than it is in temperate regions. Tropical grasses have an outstandingly high photosynthetic rate (Ludlow and Wilson 1967) and yields of 22 400 kg per hectare are easily realized provided that nitrogen, phosphorus, sulphur, and potassium nutrition are adequate. Accordingly, the main objectives are feeding value based on *in vitro* digestibility (see Chapter 16), persistence, drought resistance, and frost tolerance.

Complete apomixis, a characteristic of many tropical grasses, is a major barrier to their improvement by breeding (Hutton 1964) although the finding of sexually reproducing plants in species such as *Cenchrus ciliaris* (Bashaw 1961) has opened up possibilities for improvement by breeding in these species.

With tropical legumes, dry-matter yield is important and is related to their nitrogen and protein production per hectare as well as to their competitive ability in the pasture. Although legumes have only about half the growth rate of grasses they are more digestible and have a higher feeding value than grasses at most stages of growth (Hutton 1968*b*).

With the exception of *Desmodium intortum* and *D. uncinatum* the tropical legumes are closely self-pollinated. The *Desmodiums* have to be tripped by insects or some other agency to pollinate and the resultant seed is either selfed or crossbred.

The breeding objectives in tropical pasture legumes are

(a) increased dry-matter and protein yields;
(b) increased persistence and competitive ability with grasses;
(c) resistance to pests and diseases;
(d) improved nodulating ability;
(e) increased seed production.

8 Seeds Mixtures

Mixtures and pure stands

For the establishment of grassland swards the usual practice is to sow mixtures of grasses, generally with legumes. The advantages of mixtures as compared with pure stands are as follows.

(a) More rapid establishment of the sward and better land use.

(b) Better seasonal distribution of growth. The grazing season may be extended by the inclusion of both early- and late-maturing species.

(c) Increased production with greater palatability.

(d) Leguminous plants may be associated with the grasses to the advantage of the grasses. Legumes are richer in nitrogen and in calcium than the grasses and increase the nutritive value of the sward. A grass/legume sward is better balanced and more palatable than a pure grass stand.

When selecting species for inclusion in a seeds mixture the following points should be considered.

(1) Soil and climatic conditions. The species to be sown or planted should be indigenous or adapted to the local climatic and soil conditions. The length and severity of the dry season have an important bearing on the selection of species.

(b) The duration of the sward, i.e. whether it is intended to last for one year, two years, or three or more years.

(c) The intended use: whether intended for grazing, hay, silage, or soilage, or any combination of these.

(d) If intended for grazing, the type of stock: fattening, milking, or store cattle, sheep, pigs, or poultry.

(e) The season of the year for which the sward is intended to provide.

Seeds mixtures for long-duration or permanent pastures should normally include the following types of herbage species.

(a) Nitrogen accumulators, i.e. legumes.
(b) Rapid soil covers.
(c) Short-duration perennials.
(d) Permanent species.
(e) Bottom-filling plants.
(f) Early- and late-flowering species.

Permanent species are, of course, omitted from short-duration pastures or leys. A single grass species may meet more than one of the listed requirements, e.g. the simple mixture of perennial rye-grass and white clover of temperate grasslands meets nearly all of these requirements when mixed strains are used. Whereas in the highly developed and intensively farmed countries of temperate regions seeds mixtures have been evolved to serve specific purposes, grassland husbandry in the tropics has not yet reached this stage of development.

The contributions of legumes to pasture swards

At Palmerston North in New Zealand, Sears (1953) in a five-year trial, showed that grasses alone without manuring and with no return of dung or urine yielded annually about 2250 kg of dry matter per hectare. The addition of legumes, white and red clovers, to the sward increased the yield to 11 210 kg. In the absence of fertilizer the grasses in the clover-less sward contained 2·31 per cent of nitrogen while those in the grass/clover association contained 3·49 per cent nitrogen. It was estimated that approximately 560 kg of nitrogen per hectare was fixed annually by the clover in the grass/clover pasture. Under less favourable climatic conditions at Gore in the far south of New Zealand the nitrogen fixation by clover was only half of this figure. Under United Kingdom and United States conditions the annual nitrogen value of a clover sward is of the order of 70–220 kg nitrogen per hectare (Walker, Orchiston, and Adams 1954; Johnstone-Wallace 1937). In the United States, van Keusen and Heinemann (1958) found that the average daily live-weight gains of yearling steers were consistently higher on legume-containing pastures than on pastures without legumes, although heavier dosages of nitrogen were applied to the pastures containing grass alone. The grasses used were cocksfoot (*Dactylis glomerata*) and tall fescue

(*Festuca arundinacea*) and the legumes were lucerne (*Medicago sativa*) and ladino clover (*Trifolium repens*). In sub-tropical Queensland in Australia Miles (1949) found that the introduction of *Stylosanthes guyanensis* into a natural spear-grass (*Heteropogon contortus*) pasture contributed 130 kg of nitrogen per hectare per year. An increase in the protein content of molasses grass (*Melinis minutiflora*) due to association with tropical kudzu (*Pueraria phaseoloides*) has been reported from Puerto Rico (Vicente-Chandler, Caro-Costas, and Figarella 1953). Also in Puerto Rico Warmke (1952) found that the contribution of tropical kudzu in a Merker grass (*Pennisetum purpureum* var. *merkeri*)/tropical kudzu pasture was 205 kg nitrogen per hectare per year. In Nigeria in the humid tropics the legume *Centrosema pubescens* contributed to a star grass (*Cynodon plectostachyus*/*Centrosema* pasture 280 kg of nitrogen per hectare per year (Moore 1960).

Others grass/legume associations that have proved successful in grazing pastures in Nigeria are *Andropogon gayanus*/*Stylosanthes guyanensis*, *Panicum maximum*/*S. guyanensis*, *Melinis minutiflora*/*S. guyanensis*, *Pennisetum purpureum*/*Centrosema pubescens*, *P. purpureum*/*Pueraria phaseoloides*, and *Panicum maximum*/*P. phaseoloides*. The mixture of *Andropogon gayanus*/*Stylosanthes guyanensis* has been recommended for ley pastures in the drier Guinea savanna zone, while *Cynodon plectostachyus*/*Centrosema* has been recommended for the more humid forest and derived savanna zones. The mixtures of *Panicum* and *Pennisetum* species with *Pueraria* are intermediate in moisture requirement.

The place of legumes in tropical and sub-tropical pasture mixtures was discussed by Griffiths Davies at the 8th International Grassland Congress, 1960. '*Centrosema pubescens*, *Pueraria phaseoloides*, *Indigofera spicata* and *Leucaena leucocephala* have all been successfully and widely used with such aggressive species as Merker grass (*Pennisetum purpureum*) in Puerto Rico as have *Centrosema* with *Panicum maximum* (and *Cynodon plectostachyus*) in Nigeria and *Glycine wightii* with *Setaria sphacelata* in South Africa. In Australia successful grass/legume mixtures developed on experimental farms have included *Panicum maximum*/*Glycine wightii*; *P. maximum*/*Centrosema*; *Paspalum commersonii*/*Phaseolus lathyroides*; *Chloris gayana*/

Phaseolus atropurpureus; Paspalum plicatulum/Centrosema; P. commersonii/Desmodium uncinatum; Cenchrus ciliaris or *C. setigerus/Stylosanthes sundaica* (= *S. humilis*).

'*L. leucocephala, G. wightii, P. atropurpureus, C. pubescens, S. humilis* will maintain their place in appropriate swards under relatively heavy stocking rates of from 2·5 beasts/2 ha to 2·5 beasts/ha.'

Griffiths Davies has divided the legumes of value in tropical Australia into two groups.

(1) The strongly tropical species, which require a long, warm, and wet growing season, e.g. *Pueraria phaseoloides, Calapogonium mucunoides, Centrosema pubescens, Leucaena leucocephala, Glycine wightii;* and

(2) a group that can survive dry, even arid, conditions for several consecutive months, e.g. *Stylosanthes guyanensis, S. humilis, Phaseolus lathyroides, P. atropurpureus, G. wightii, Desmodium uncinatum, D. sandwicense, D. scorpiurus, Vigna repens, Indigofera spicata, Lotononis bainesii*, and *Dolichos lablab*.

Few species are frost-resistant and all except *G. wightii* (certain strains) and *Lotononis bainesii* are entirely defoliated by one or two hours at temperatures below 0 °C. The tropics and sub-tropics of Australia are arid with the exception of the coastal areas and frosts are prevalent. Annual rainfalls vary from 25 cm to 100 cm but there are narrow bands with rainfalls of 100–250 cm along the east and north coasts.

Nitrogen deficiency is widespread in the grassland areas of tropical Australia but nitrogenous fertilizers are uneconomic under extensive conditions. The key to increased production is the successful introduction of persistent pasture legumes adapted to the local environment, compatible with grasses, and able to withstand grazing. While no single legume has yet been found to satisfy all these conditions, several have shown considerable promise. *Desmodium uncinatum, Lotononis bainesii, Indigofera spicata, Glycine wightii, Centrosema pubescens*, and *Phaseolus lathyroides* have all greatly improved the yield of dry matter and protein per hectare, and the stock-carrying capacity, when grown in association with appropriate grasses.

The mode of transfer of nitrogen from legume to associated grass has been studied by Whitney and Kanehiro (1967) in pot-culture experiments using *Desmodium intortum*, *D. canum*, and *Centrosema pubescens* with pangola grass as acceptor. The combined effects of root losses, leaching of leaves, and leaf fall provided an adequate explanation for nitrogen transfer. In field trials in Hawaii the same author (1968) observed that transfer of nitrogen to the grass was not more than about 5 per cent for *D. intortum* and 11 per cent for *C. pubescens*.

Some tropical seed mixtures

Little is known concerning seeds mixtures in the tropics and knowledge necessary for the compounding of mixtures is only now being collected. Some successful seeds mixtures are recorded on p. 56.

The following pasture mixtures have been used successfully on the tropical coast of Queensland.

Species	Rate (kg/ha)
Pangola	Sprigs
Centro	2
Stylo	1
Tropical kudzu	1
Para	Sprigs
Centro	4·5
Tropical kudzu	1
Guinea	2–3
Centro	2
Tropical kudzu	1
Siratro	1
Stylo	1
Rhodes	3–4·5
Siratro	1
Guinea	3
Molasses	1–2
Stylo	1–2

Rhodes grass/lucerne. 11–16 kg per hectare Rhodes grass and 2 kg per hectare lucerne is recommended in East Africa. Pasture should be well established before grazing.

Rhodes grass/Clitorea ternatea. 11kg per hectare *Clitorea* drilled in rows 60 cm apart and Rhodes grass then broadcast (Bunting and Lea 1956). Good grazing pasture in Sudan central rainlands.

Molasses grass/tropical kudzu. 4–6 kg/ha molasses grass and 6–7 kg/ ha tropical kudzu in Colombia (Vasquez 1957).

Molasses grass/Centrosema. 2–3 kg/ha molasses grass and 2–3 kg/ha *Centrosema.*

Guinea grass/Centrosema. 4–6 kg/ha Guinea grass and 2–3 kg/ha *Centrosema.*

<table>
<tr><td>Elephant grass/Centrosema</td><td rowspan="3">Cuttings planted at 60–90 cm in rows 90–120 cm. apart plus Centrosema at 2–3 kg/ha (Walsh 1959).</td></tr>
<tr><td>Pangola grass/Centrosema</td></tr>
<tr><td>Para grass/Centrosema</td></tr>
</table>

9 Establishment of the Sward

THE type of pasture produced is governed by several factors: the grass and legume species sown or planted, the soil fertility, and the degree of success achieved in the establishment of the sward.

In temperate climates

The procedure for the establishment of temperate-region pastures is more or less standardized. The steps involved are preparation of the land, determination of the time of sowing. and the method of sowing, the application of fertilizers, and deciding how soon to graze or harvest the sward. Clearing of the land is seldom necessary although levelling and the provision of adequate drainage may be desirable.

The land is ploughed, preferably some months before sowing of the seed, and well cultivated by repeated harrowing and ring-rolling in order to achieve a fine, even tilth and a firm seed-bed. Large quantities of lime (5–7½ tons ground limestone per hectare in Ireland and Wales) are normally applied before sowing, and the major nutrients, nitrogen, phosphate, and potash, at the time of sowing. A grain drill with fertilizer attachment can be used conveniently for supplying fertilizer with the seed. In New Zealand nitrogen and potash are rarely applied from the manure bag, potassium being returned to the sward in the urine of the grazing animal and nitrogen fixation by the legume (clover) supplying the nitrogen essential to the growth of the grass. It is common practice to apply 400 kg of superphosphate per hectare at sowing and again six months after sowing. An annual application of 400 kg per hectare of superphosphate maintains the clover, which has a high phosphate requirement. Localized deficiencies of other elements, e.g. magnesium, molybdenum, cobalt, may also be supplied at sowing when their deficiency is indicated by the health of the grazing animal or the appearance of the sward.

The time of sowing is dependent upon the attainment of adequate conditions of soil and climate for germination of the seed. In temperate regions early spring is generally the most suitable time.

Several methods are available for sowing. The seed may be broadcast by hand, mechanical broadcaster, or aeroplane, or may be drilled. As the seed is small, shallow seeding is important, but for satisfactory germination it should be covered with soil. Broadcast seed may be covered by light harrowing or discing, or a 'cultipacker' may be used. The latter machine consists of two ring-rollers arranged in tandem. The seed is broadcast from a seed-box immediately behind the first roller and falls mainly between the ridges. These are split over the seed by the second roller.

The establishment of the sward is greatly assisted by consolidation of the ground which is most economically achieved by the trampling of grazing animals. It is generally considered that on adequately drained land stock should be turned on to the sward as soon as there is enough herbage to feed them. A short-duration, early-maturing grass or cereal crop is sometimes inter-sown as a 'nurse' or companion crop which will provide feed before the main grass crop is ready for grazing. The companion crop has the disadvantages of competition with the main grass crop for soil moisture and nutrients and an undesirable shading effect. Its use is no longer generally recommended in temperate regions. In the tropics, however, *Chloris* and *Melinis* have been established successfully under maize as companion crop (in East Africa). In semi-arid or arid regions, where the water supply is limited, a companion crop should never be grown. Grazing during the first season should be on an 'on-and-off' basis with recovery periods of at least three weeks between grazings.

In the tropics

While the general principles of pasture establishment are universal, minor modifications are essential in order to achieve success under tropical conditions. The first stage in the establishment of a tropical pasture is almost invariably clearing of the land. Frequently this involves the felling of trees and stumping and may cost £100 per hectare (Ibadan, Nigeria, 1961). Wherever practicable fire is the principal means of bush clearing. Some of the cost of clearing may be recovered by growing a cereal cash crop prior to sowing the grass crop.

Although tropical and sub-tropical soils are often acid in reaction, lime is seldom applied and, indeed, the response to lime shown by many arable crops in the tropics appears to be of little significance (Nye 1952) and in Kenya and Rhodesia lime has not assisted the establishment of grass leys. Many tropical soils fix phosphate readily and crops show little response to phosphate. Responses to superphosphate obtained with groundnuts in Northern Nigeria may be due to the sulphur content of the fertilizer. However, significant responses to superphosphate have been shown by grass/legume leys in southern Nigeria and its application appears to be desirable. The most common deficiency in tropical soils is nitrogen and this nutrient, generally as ammonium sulphate or urea, is almost invariably applied with advantage at sowing. Deficiency of potash militates against the successful establishment of legumes in a grass/legume pasture but it is generally present in sufficient amounts in tropical soils and is not normally supplemented. Minor elements are not generally supplied and organic manures are seldom available in sufficient quantities.

In preparation of the seed-bed the land, after clearing and stumping, may be ploughed and sometimes crossploughed with a disc-plough, disc-harrowed, and then given a moderately good tilth with a cultivator. It is not usual to prepare the seed-bed with the same degree of fineness of tilth as in temperate regions, nor is this always desirable. A rougher soil surface gives a measure of protection against wind and rain erosion and helps to reduce evaporation. For propagation by means of cuttings and rhizomes a fine seed-bed is not necessary.

Grass-killing chemicals may be used to destroy old turf before preparing a seed-bed by surface treatment. Bipyridylium herbicide is non-toxic in soil but is rapidly absorbed into foliage. An interval of three weeks between spraying and sowing is recommended.

The best time of sowing is in the early rains and the normal methods of temperate regions described above may be used. Seed rates vary with the species and the conditions. (Seed rates adopted for several species are given in Chapter 4.) It is normal to adopt a low seed rate in unfavourable soil and climatic conditions in order to reduce competition between the seedlings for moisture and nutrients.

For the vegetative propagation of tropical grasses, stem cuttings (e.g. Guinea and elephant grass) may be planted in the same way as sugar-cane setts. With rhizomatous or stoloniferous species (e.g. star grass) the pieces of rhizome or stolon used as planting material may be broadcast over the field and disced in. For success of this method there must be rain within two days. The method is not suitable in the drier parts of the tropics, where the planting material should be dropped in furrows and covered with soil either by hand (or foot) or mechanically.

The spacing adopted for vegetative propagation varies with the type of grass. A wider spacing is satisfactory for spreading stoloniferous or rhizomatous species that quickly cover the ground, while closer spacing is required for tall, stool-forming species. Paterson (1938) recommended a spacing of 90 cm × 60 cm for elephant grass and 60 cm × 45 cm for Guatemala grass, both stool-forming grasses. For the stoloniferous grasses, a usual spacing is in rows 90 cm apart with 45–60 cm intervals within the rows. In humid tropical areas or under irrigation, close spacing gives the highest yields, whereas in the more arid areas a wider spacing is preferable to reduce plant competition and to allow adequate development of the root system under less favourable conditions.

Early grazing assists the establishment of stoloniferous grasses; the trampling by the animals pushes the stolons into the ground and aids rooting. Stool-forming grasses are easily uprooted and should not be grazed until the root systems have become well developed. In either case grazing should be on an 'on-and-of' system with recovery periods. In semi-arid areas stock are excluded from the sward until the second season.

Tropical legume seed should be treated with the appropriate inoculum before sowing. Lime-pelleting may assist nodulation. Superphosphate and other acid fertilizers will destroy the inoculum if seed and fertilizer are mixed together. In these circumstances lime-pelleting is necessary. Treatment of legume seed with Dieldrin is a useful control against bean fly.

10 Management and Manuring

THE aim of pasture management for animal production is to secure the maximum quantity of nutritious and digestible animal food distributed as widely as possible over the grazing season and to ensure the efficient utilization of the food produced. This generally involves conservation of surplus food produced at the height of the growing season for use during periods of reduced growth, e.g. in the dry or cold seasons.

Controllable factors in management are dealt with under the following main headings.

MECHANICAL TREATMENT

This involves harrowing and mowing. The fields may be lightly harrowed at regular intervals to spread animal droppings. These droppings are more easily broken up after rain. The rank growth that is produced in proximity to the droppings is unpalatable to cattle. Harrowing is also advantageous to break up matted vegetation and tangled stolons.

Mowing is an important adjunct of grazing in the maintenance of good pasture swards. By prevention of flowering and by the encouragement of tillering it increases productivity. Coarse, unpalatable growth left by the grazing animal should be removed by the mower. Mowing is an effective means of weed control but unfortunately is often impracticable in tropical pastures because of the presence of rocky outcrops and hidden stumps.

In arid regions subject to high-intensity rainfall for short periods, the run-off may be effectively reduced by making small pits in the soil by means of an eccentric disc with adjacent discs set off-centre in different directions. The pits collect the run-off and assist water retention by the soil (Rauzi and Lang 1956).

THE USE OF FERTILIZERS

Mineral fertilizers

Since the more productive pasture grasses demand high soil fertility, the regular use of mineral fertilizers is necessary to maintain a high level of production.

The four major nutrient elements required are nitrogen, phosphorus, potassium, and calcium, while magnesium, sulphur, manganese, zinc, copper, boron, molybdenum, and cobalt may be deficient in certain areas. The effect of the minor elements is greatest on the legumes. Copper is needed for reproduction of the plant, zinc for early vegetative growth, and molybdenum for nitrogen fixation by the root-nodule bacteria.

The major elements

In established pastures the requirements for the major elements vary with the use of the pasture, e.g. beef cattle return most of the phosphorus and potassium to the soil in their urine whereas milking cows cause appreciable losses of these elements. The tables on p. 63 show the return of nutrients to the soil in dung and urine and the losses incurred when animal produce is sold off the farm.

Calcium

Herbage plants, especially legumes, have a high lime requirement. Lime is generally applied in temperate regions to produce a satisfactory soil reaction rather than to increase yields. The soil reaction for the better grasses and clovers is pH 5·5 to 6·5. Lime flocculates the clay fraction and improves soil structure. It reduces phosphate fixation and increases the activity of desirable micro-organisms. Excess lime in the soil lowers the availability of important minor elements, such as iron, manganese, copper, zinc, and boron.

In the tropics lime is looked upon as a fertilizer rather than as a soil conditioner and heavy dosages may be harmful. Application of lime should always be based on field trials. Tropical soils often have low buffering capacity and hence the application of a small amount of lime may cause a considerable rise in pH. In preliminary field trials at Fashola in southern Nigeria grasses (*Cynodon*, *Axonopus*, and

Panicum) gave no response to lime. The legume *Centrosema pubescens* gave no response to lime when applied singly but gave a greater response to LPK (lime + phosphate + potash) than to phosphate or potash singly or in admixture. It was concluded that when lime was applied to a legume or grass/legume sward phosphate and potash should also be applied.

TABLE 3

Calculated quantities (kg per hectare) of common fertilizers equivalent to dung and urine returned annually to the soil by animals grazing on productive New Zealand pasture (Sears 1950)

	Sulphate of ammonia (kg)	Super-phosphate (kg)	Carbonate of lime (kg)	30% potash salts (kg)
Urine	2236	Trace	Trace	1870
Dung	821	765	300	221
Total	3057	765	300	2091

TABLE 4

Annual fertility drain per hectare incurred by disposal of animal produce or carcases (Sears 1950)

	Annual loss per hectare in fertilizer ingredients			
	Sulphate of ammonia (kg)	Super-phosphate (kg)	Carbonate of lime (kg)	30% potash salts (kg)
Milking cows at 6825 litres milk/ha sold off farm	179	67	24	34
Cattle, 2·5 beasts (454 kg)/ha raised and fattened on farm	130	86	45	6·2
Store cattle or sheep (fattened only on farm)	Losses negligible			
Sheep at 15/ha, 68 kg live-weight each, raised and fattened on farm	117	54	24	54

Nitrogen

Although nitrogen is the principal nutrient required for the continued growth of grass the usual tendency is to rely upon nitrogen fixation by the root nodules of the legumes in a grass/legume pasture rather than on the manure bag. Nitrogen fertilizer is, however, applied in temperate regions to extend the grazing season. When applied early in the season it can increase production by one-third. Late applications may also be justified. A customary dosage for this purpose is 250 kg of ammonium sulphate per hectare. When nitrogen is applied phosphate and potash should also be applied.

Suppression of the legume by nitrogen

Nitrogen fertilization of a grass/legume sward by stimulating the growth of the grasses reduces the legume content of the sward. Suppression of the legume is due to the effect of shading by the grass, and increased competition for rooting space, nutrients, and soil moisture. The reduction in the quantity of nitrogen contributed by the legume must be compensated by the nitrogen fertilizer before any increase in the nitrogen content of the pasture can be observed. Application of less than 125 kg of nitrogen per hectare to a clover sward depresses the yield of protein ($N \times 6{\cdot}25$). Higher levels of nitrogen fertilizer show an increase in protein production.

Nitrogen fertilizers increase the total nitrogen content of grasses. Heavy nitrogen dressings increased the nitrate N in stands of pure rye-grass and of permanent pasture (Nowakowski 1958).

Nitrogen fertilization of tropical grasses

Legumes are likely to remain the major source of nitrogen for tropical pastures. The advantages of grass/legume pastures are low-cost animal production and adaptability to extensive and semi-intensive grazing systems. When fertilizer nitrogen is available at a relatively low cost, owing to increasing industrialization, and where land is limiting there is a tendency to replace the legume by the fertilizer bag. Nitrogen is a limiting factor for grass growth, especially early in the growing season. The later commencement of active growth of legumes as compared with grasses and the low persistence of certain tropical pasture legumes may prevent an associated grass from attaining its potential yield, because of inadequate nitrogen nutrition.

A grazing trial was conducted over four years in Brazil (Aronovich, Serpa, and Ribeiro 1970) to compare commercial fertilizer and *Centrosema pubescens* as sources of nitrogen for fertilization of pangola grass for beef production. The treatments were

(a) pure stand of pangola grass fertilized with 100 kg per hectare per year of nitrogen (as calcium nitrate);

(b) mixed pasture of pangola grass and *Centrosema pubescens* with legume content ranging from 25 to 50 per cent;

(c) pure stand of pangola grass without nitrogen fertilization as control.

Live-weight gain per hectare was highest with commercial fertilizer (more than 50 per cent above control). The mixed grass/legume pasture produced 17 per cent more than the control.

Outstandingly high animal production resulting from the heavy application of nitrogen to grass has been reported in recent years. In the sub-tropical coastal lowlands of south-east Queensland T. Evans (1970) achieved consistent live-weight gains greater than 1200 kg per hectare per year on pangola grass fertilized with 448 kg of nitrogen per hectare when grazed at a stocking rate of 7·4 yearling beef heifers per hectare. At Parada in northern Queensland J. Evans (reported by Hutton 1968) achieved 2000 kg per hectare per year live-weight gain on irrigated pangola grass heavily fertilized with nitrogen. Salette (1970) at Guadeloupe in French West Indies obtained a maximum yield of more than 54 tons of dry matter per hectare from pangola grass corresponding to more than 800 kg of nitrogen per hectare in the crop. Over 80 per cent of the nitrogen was recovered in the crop.

It is unlikely that all species will be equally responsive to nitrogen. Other grasses found to respond in Queensland are kikuyu grass, *Setaria sphacelata*, and a new perennial *Sorghum* hybrid.

The influence of heavy applications of nitrogen fertilizer on the pasture and on the animal was investigated by Hodgson and Spedding (1966) (calves) and Large and Spedding (1966) (lambs). The total quantities of nitrogen applied to the pasture ranged from 185 to 1100 kg per hectare per year. The heavy nitrogen treatments increased the concentrations of total N, N.P.N., and nitrate N in the herbage, depressed the concentration of soluble carbohydrate, and slightly increased the digestibility. There were no adverse effects on the health of the animals attributable to the high nitrogen.

Phosphate and potash

In grass/legume swards phosphate and potash are required especially by the legume, which supplies nitrogen, fixed by its root-nodules, to the associated grasses. Potassium is returned in the urine of the grazing animal and additional potassium is not generally required on well-established pastures.

In the United Kingdom, after an initial heavy dressing of superphosphate at sowing, annual dressings of 250 kg of superphosphate per hectare are usual. The New Zealand practice is to apply 400 kg superphosphate per hectare per year. Small dressings applied once or twice a year are generally more effective than large dressings at less frequent intervals. In some soils, e.g. in Rhodesia, the fertilizer is ploughed in or it may be banded with the seed. Deep application is necessary to get the best response and to avoid fixation at the surface. On soils of high phosphate-fixing power, basic slag and rock phosphate have often proved superior to superphosphate.

Phosphate and potash used for grazed leys at Rothamsted (1959) in ley-fertility experiments and thought to be sufficient but not 'luxury intake' were P_2O_5 37·6 kg per hectare (188 kg superphosphate), K_2O 75·3 kg per hectare (150 kg potassium sulphate).

Increasing the supply of phosphate fertilizer increases the phosphorus content of the herbage. The first effect of the application of potassium is to increase the yield. If the potassium supply is increased beyond that required for a full crop then the potassium content of the herbage increases without a corresponding increase in yield. Lehr (1951) showed that increasing the potassium intake reduces the intake of calcium, magnesium, and sodium. Kemp (1958) has shown that high potassium in grass is a factor that may lead to increase in grass tetany. Reduction in the magnesium level may also lead to hypomagnesemia in cattle. It is recommended that the minimum amount of potassium consistent with obtaining the desired crop yield and preventing depletion of soil reserves should be applied.

Solid and liquid animal manure

Liquid manure is used to a considerable extent on some of the pastures of north-west Europe. Being low in phosphorus and high in potassium it needs to be balanced by dressings of superphosphate. Heavy dressings of liquid manure in spring may lead to grass tetany owing to excess potassium and depletion of calcium and magnesium. Solid manure is seldom applied except during establishment and on permanent pasture used for hay.

In investigations in New Zealand it has been shown that the return to the pasture of dung and urine separately increased the yield of mixed pasture 15 per cent and 18 per cent respectively, while the return of both increased the yield 32 per cent. These applications of urine and dung also increased the percentage of grass in relation to clover (legume). Taking grass alone, the yield was almost doubled by returning the dung and urine. No response to superphosphate or lime was obtained in pastures receiving the normal return of dung and urine (Sears 1953).

GRAZING MANAGEMENT

The objectives of grazing management are

(a) to maintain high production of good-quality forage for the longest possible period;

(b) to maintain a favourable balance between herbage species;

(c) to achieve efficient utilization of the forage produced; and

(d) high animal production.

Although maximum yields of dry matter are obtained when pasture is harvested at or near maturity, the nutritive value and digestibility of the forage at this advanced stage are low. Young growth with a high proportion of leaf to stem is of the highest quality with maximum content of crude protein and minimum crude fibre. Management should be regulated to prevent most of the plants from flowering and, if this cannot be achieved by grazing, the mower should be used. Unconsumed forage removed by mowing may be conserved as silage or dried grass.

Animals thrive best when allowed 'full bite' and no more. On temperate pastures this is considered to be 15 cm for mature cattle and less than half this height for calves.

Young, flush herbage grown rapidly under conditions of high soil fertility when intensively grazed may induce scouring and bloat in the grazing animal. The non-protein nitrogen of young flush pasture may exceed 50 per cent of the total nitrogen. Sullivan of the Pasture Research Laboratory in Pennsylvania considers that when the non-protein nitrogen content exceeds 20 per cent digestive disturbances will result. Bloat is generally associated with a high proportion of legumes in the pasture mixture. When the legume content does not exceed 50 per cent the condition is unlikely to occur. The legume content may be kept down to a safe proportion by application of nitrogen and withholding of phosphate. Spraying of susceptible pastures with mineral oil as anti-foaming agent has proved effective in bloat prevention. Bloat on tropical pastures is in rare occurrence.

Fairly close grazing promotes tillering, but too frequent defoliation, especially in the early stages of growth, retards growth and root development, leads to diminution in vigour, and may lead to replacement of the more palatable and desirable species in the sward by poor-quality species that escape grazing, and weeds (Hunt 1952).

Undergrazing or lax grazing, on the other hand, leads to diminution in the feeding value of the pasture. With increase in maturity of the herbage the crude protein content falls, crude fibre content rises, and nutritive value and digestibility fall until the forage becomes no better than a maintenance ration. Lax grazing is essential for the persistence of tall stool-forming grass species whereas prostrate stoloniferous grasses are able to withstand close-grazing. In order to preserve a favourable balance of species careful control of grazing is required.

Systems of grazing

Continuous grazing

This is an extensive system of grazing in which the stock remains on the same pasture area for prolonged periods. The stocking rate is relatively low. Continuous grazing is the normal practice on range lands and tropical grasslands where fencing is economically impracticable. While this system under low stocking rates may prove no less productive than rotational grazing, tropical grasslands are normally undergrazed during the rains and overgrazed in the dry season with consequent deterioration of the sward. Amelioration of these conditions by reduction of the stocking rate during the dry season when pasture growth is slow and feeding value low is seldom feasible owing to the absence of a market at that time. The animals are generally carried over the dry season on a submaintenance ration.

A serious disadvantage of continuous grazing in the tropics is the build-up of tick and nematode infestation. If young stock are allowed to graze continuously on the same pasture as older stock they become heavily infested with helminth parasites and their growth is retarded. Under a system of rotational grazing the degree of infestation is considerably reduced.

Rotational grazing

Rotational grazing is an intensive system of grassland management practised on improved permanent or ley pastures. It overcomes the disadvantages of undergrazing and overgrazing. The grazing area is subdivided into a number of paddocks, usually at least six, and the animals are moved systematically from one to another of these in rotation. The stocking rate on each paddock is high, for instance 25 cow-equivalents per hectare. Each paddock is grazed for a period of three to seven days, the length of the grazing period depending on stocking rate and herbage growth rate, after which the stock are moved on to the next paddock while the first is rested. By the time the last paddock in the sequence has been grazed down the first paddock should be ready for grazing for the second time. The aim of this system of grazing is to use the pasture when it is young and highly nutritious and then to allow an adequate recovery period.

In order to ensure the most efficient utilization of the pasture the grazing animals may be divided into two groups, high producers (dairy cows and fattening cattle) and low producers (dry cows and stores). The high producers enter the paddocks first for a short period to graze off the 'cream' and are then followed by the low producers who consume the remainder of the pasture. Alternatively, high-producing cattle may be followed by sheep.

As grass growth is seasonal, periods of high production will alternate with periods of low production and it is necessary either to change the stocking rate to meet these conditions or to remove surplus pasture by the mower for conservation and later use as hay, silage, or dried grass.

The principles and practice of rotational grazing are lucidly described by Voisin (1959) in his book *Grass productivity*. Voisin formulated four laws of rotational grazing.

First law. Before a sward, sheared with the animals' teeth, can achieve its maximum productivity, sufficient interval must have elapsed between two successive shearings to allow the grass

(a) to accumulate in its roots the reserves necessary for a vigorous spurt of regrowth;

(b) to produce its 'blaze of growth' (or high daily yield per acre).

Corollary of first law. The rest period between two successive shearings varies with the season, climatic conditions, and other environmental factors.

Woodman in England considered that an interval of one month should elapse between successive close grazings, the interval being extended to five to six weeks during dry periods. Voisin in France allows an optimum period of eighteen days at the time of rapid growth (May) and twice this in the dry period of slower growth (August; late summer in France).

Second law. The total occupation period on one paddock should be sufficiently short for a grass sheared on the first day (or at the beginning) of occupation not to be cut again by the teeth of these animals before they leave the paddock.

The period of occupation should not exceed four days, at the very maximum six days.

Third law. The animals with the greatest nutritional requirements must be helped to harvest the greatest quantity of grass of the best possible quality.

First corollary of third law. Grass with an average height of 15 cm in the case of permanent pastures (and of at least 23 cm in temporary grazings) allows the cow to harvest maximum quantities of high-quality grass. (Note that in the tropics the preferred height of grazing in grass/legume pastures is 38–46 cm.)

Second corollary of third law: The less scraping (or finishing-off) imposed on the cow, the more grass she harvests.

Fourth law. If a cow is to give regular milk yields she must not stay longer than three days on the same paddock. Yields will be at their maximum if the cow stays on one paddock for only one day.

By application of the last two laws Voisin considers that the individual yield of an animal will be increased by 20 to 30 per cent. Obeying the first and second laws will, at least, double the yield of grass. Voisin's rotationally grazed pastures produced 7123 kg starch equivalent per hectare in 1954.

In order to compensate for seasonal fluctuations in grass growth it is desirable to reduce the grazing intensity during the period of minimum regrowth (late summer in France). As an alternative to the sale of surplus animals Voisin suggests the following.

(a) Disengaging of paddocks that were shut up for hay or silage when grass growth was most rapid. These paddocks should be brought back into the rotation.

(b) Application of nitrogen to extend the grazing season.

(c) Feeding supplements or soilage or grazing a temporary ley.

Rotational creep grazing

Rotational creep grazing for lambs while still on the ewes has made possible the very high live-weight production of 900 kg per hectare per year in the north of England. In this modification of rotational grazing the lambs are allowed to graze new pasture ahead of the ewes whose movement is controlled by fencing.

Influence of stocking rate

McMeekan (1960) has repeatedly emphasized that the full benefits of rotational (controlled) grazing may not be realized unless the system is associated with a high stocking rate to exploit fully the efficiency of the more intensive grazing method. It has been demonstrated that increased stocking rate under continuous grazing (set stocking) is not accompanied by as high an increase in output per hectare as under rotational (controlled) grazing. Under both grazing systems increased stocking rate leads to reduced production per animal but increased production per hectare.

Strip grazing (*close-folding, ration grazing*)

Strip grazing is a more intensive method of rotational grazing based on the use of the electric fence. A movable electric fence is placed across a grazing paddock and is moved forward once or twice each day. The amount of herbage offered to the animal is thus restricted, selective grazing is minimized—the pasture being uniformly consumed—and there is less wastage by trampling and defecation. In order to prevent grazing of the regrowth a second fence may be brought up behind the first one. The fences are moved progressively across the paddock.

Obviously, the strip-grazing technique will be of value only on highly productive nutritious pastures. Any restriction of intake on poor-quality pastures will lead to a fall in animal production. The technique is mainly applied to high-producing dairy pastures.

Increases of the order of 15–20 per cent in the efficiency of utilization of high-quality pasture by dairy cattle with strip grazing as compared to rotational grazing have been reported by several workers in Western Europe, e.g. de Geus and t'Hart (1952); Holmes (1952). In New Zealand, on the other hand, no appreciable difference in the productivity by the two methods was observed (McMeekan and Hancock 1954; Wallace 1955 and, more recently in Victoria, Australia, Freer 1959), on irrigated dairy pastures, found no significant difference in animal production between strip and rotational grazing. The average weight of pasture nutrients utilized per year by the rotationally grazed group was 6475 kg starch equivalents per hectare and by the strip-grazed group 6486 kg starch equivalents per hectare.

Deferred grazing

As the name implies, deferred grazing is the setting aside of certain pasture paddocks for use at a later stage, e.g. the practice of conserving 'standing hay' in the tropics. With maturity the feeding value and palatability of grass fall off and generally little more than a maintenance ration is provided. Even this will be of great value at the peak of the dry season when range pastures are grazed almost bare. The leafier grasses are the most suitable for standing hay. Giant star grass (*Cynodon plectostachyus*) and Guinea grass (*Panicum maximum*) have been used successfully for this purpose in Nigeria. In temperate regions deferred grazing may be adopted in order to provide winter feed. Supplementary winter pastures of cocksfoot and lucerne are commonly used in Great Britain and brome grass and lucerne in the United States.

The practice is also used as a means of improving range pasture. By allowing the plants to mature before grazing their vigour is built up, root systems are allowed to develop, and self-sown seedlings to become established. Low-growing legumes in the sward will, however, be suppressed by the taller grasses and the balance of species will be changed.

Soiling or zero-grazing (Watson and Runcie 1960)

Soiling, or the feeding of cut crops (including grass) to housed stock, has been described as zero-grazing, no doubt because the animal can be reared on grass without the necessity of grazing the grass. It is not a new practice but an old one that has been revived owing to the reduction in the costs of cutting and carting made possible by the evolution of modern machinery, in particular the forage harvester.

The advantages of soiling as compared with rotational grazing are higher animal production owing to more efficient herbage utilization; there are no losses from fouling and trampling and the pasture may be harvested at a more advanced stage of growth. Herbage intake is uniform. A greater measure of bloat control is possible especially if the pasture is allowed to wilt a little before feeding.

Disadvantages are the cost of equipment, which restricts the practise to herds of at least thirty cows, and the need to provide bedding for the housed animals. Manure disposal may present a problem.

Soiling permits an increased stocking rate and an intensification of management. It dispenses with the need for fences and piped water supplies and the animals are protected from heat and flies.

At the University of Khartoum's farm at Shambat soiling has been practised successfully for many years. The cattle, of Zebu origin, are kept in pens or in a yard except for a brief exercise in the cool of the morning. The principal soilage is berseem (*Trifolium alexandrinum*).

The desert area on which the farm is located has few grazing grasses except after the rains and these are short-lived. The practice should be advantageous in semi-arid or arid parts of the tropics where forage for cutting and feeding to the stock can be grown under irrigation.

At the University farm, Ibadan, Guatemala grass (*Tripsacum laxum*) has been used extensively for soiling.

EFFECT OF MANAGEMENT ON BOTANICAL COMPOSITION

The botanical composition of a sward is largely determined by the management. Thus close grazing early in the grazing season followed by a period of rest will tend to suppress early-maturing species and favour the later starters. Deferring grazing until later in the season will have the reverse effect.

In a five-year trial at Palmerston North in New Zealand, Sears (1953) studied the effect of superphosphate and dung and urine on the botanical composition of a rye-grass/clover pasture. Without dung and urine the grass/clover mixtures were dominated by the clovers (mainly white clover) but where dung and urine were returned the grasses were dominant. Superphosphate gave responses in growth but only in the grass/clover mixtures not receiving dung and urine. No response to superphosphate was obtained in any cloverless pasture or in the grass/clover mixtures receiving the normal return of dung and urine from the mature wethers used in the trial. Red-clover growth was maintained only in the unfertilized mixtures under mowing and not receiving dung and urine. Red-clover growth was adversely affected by grazing, by urine, and by superphosphate, either directly or through resultant white-clover competition.

The effect of continuous as compared with rotational grazing on a New Zealand pasture sward is illustrated by the results of a grazing trial quoted by Levy (1955). Close continuous grazing resulted in a sward dominated by perennial rye-grass and white clover, with production mainly in spring and early summer; under a system of rotational grazing with long rest periods there was more short rotation than perennial rye-grass, and red clover made a large contribution to the yield in summer and autumn, production during this period being nearly double that from the continuously grazed sward. Under rotational grazing with a short rest period a sward intermediate in character was obtained.

Tables 5–7 illustrate the effect of grazing management on botanical composition of pastures in the United Kingdom.

TABLE 5

The effect of management on botanical composition of a sward (from Jones 1933)

	Proportion of species present at end of second year			
	Perennial rye-grass	Rough-stalked meadow grass	Crested dogs-tail	Wild white clover
Not rested	40·0	2·9	0·6	54·8
Rested in May	32·2	7·2	0·2	60·2
Rested from Feb. to April	45·7	2·5	2·7	47·2
Rested from Nov. to April	54·8	1·3	1·3	40·7

TABLE 6

Botanical analysis in October 1947 of pastures under three systems of grazing management

	Type of management		
Species	Continuous grazing (%)	Short-spell rotational grazing (%)	Long-spell rotational grazing (%)
Perennial rye-grass	43·3	41·6	31·1
Short-rotation rye-grass	2·9	21·7	40·2
Timothy	6·8	1·2	—
Cocksfoot	1·0	3·2	—
Other grasses	—	4·0	0·7
White clover	42·4	24·3	23·3
Red clover	2·8	2·8	4·2
Other species	0·8	1·2	0·5
	100·0	100·0	100·0

TABLE 7

Botanical analysis and yield during summer and autumn of pastures under three systems of grazing management

Species	Type of management Continuous grazing (%)	 Short-spell rotational grazing (%)	 Long-spell rotational grazing (%)
Rye-grass (short rotation and perennial)	37·3	48·8	20·4
Timothy	—	0·5	Trace
Cocksfoot	0·6	1·0	0·8
Other grasses	1·6	3·5	Trace
White clover	54·7	31·8	21·4
Red clover	5·1	12·1	55·3
Other species	0·7	2·3	2·1
	100·0	100·0	100·0
Yield for period (in kg of dry matter/hectare)	3250	3810	5710
Crude protein (as % of total dry matter)	25·4	23·8	20·4
Crude fibre (as % of total dry matter)	15·5	17·3	19·9
Number of grazings	Continuous	5	2

Continuous versus *rotational grazing*

One of the most controversial issues in grazing management is the relative merits of continuous and rotational (intermittent) grazing. There are advantages in both systems, and good management may be best served by one or the other depending on the prevailing circumstances. To claim that rotational grazing invariably leads to increased animal production is quite unjustified. Greater returns attributed to subdivision and rotational grazing are often merely a reflection of increased stock numbers and closer supervision. It is not to be expected that rotational grazing will show any economic advantage except at fairly high stocking rates.

The arguments in favour of a definite interval between defoliations, i.e. a period of rest from grazing, are not entirely convincing. Normally, defoliation slows down all aspects of growth but it has been shown by Mitchell and Coles (1955), working with perennial ryegrass in New Zealand, that low light intensities at the base of the grass plant inhibit tillering, so that increased light intensity brought about by defoliation may lead to increased tiller production. On the other hand, an adequate rest period after grazing allows root recovery, without which lucerne stands rapidly die out (Peart 1968). Many of the tropical legumes cannot tolerate hard grazing. A system of rotational grazing is essential for their persistence.

The effect of grazing frequency on a kikuyu grass/*Glycine wightii* pasture in presence of nitrogen fertilizer was investigated at Wollongbar in New South Wales, Australia (R.L. Colman 1968). In the first year of the experiment, after application in the spring of up to 244 kg of nitrogen (as ammonium nitrate) per hectare, there was a marked reduction in legume yield with four-weekly grazing but little effect with eight- or twelve-weekly grazing. In the absence of nitrogen the effect of increased grazing frequency in reducing legume yield was less than when nitrogen was applied. In the second year, legume yields declined further with four-weekly and eight-weekly grazing but there was little effect with twelve-weekly grazing except at the highest nitrogen rate. In experiments at Wollongbar with kikuyu grass alone it was found that an increase in frequency of defoliation from once every twelve weeks to once every two weeks gave an increase in nitrogen content of the dry matter from 1·3 to 3·0 per cent but a reduction in yield of 34 per cent.

In the tropics, internal and external parasites assume much greater importance than in temperate regions. It has been established that tick control is achieved by use of rotational grazing (Dunwell 1967). Reduced worm infestation and consequent better weight gains have been reported by Dickson (1959) and Spedding and Large (1959). On the other hand, continuously grazed (set-stocked) animals do not suffer nutritional scours which may be induced by rapid changes of feed.

As pasture species in the wet tropics mature very rapidly, it is desirable to graze pastures in these regions as rapidly as possible and at frequent intervals. The stocking rate should be heavy so that a large number of cattle are concentrated on a small area for a short time and are rotated around the pastures at frequent intervals. It may be economic to conserve forage from the wet-season flush for feeding out during the dry season when pasture availability is low.

11 Grassland Utilization

Seasonal influences

THE best utilization of grassland is obtained when the seasonal rise and fall in grass production is made to coincide as far as possible with the nutrient requirements of the grazing animal during its cycle of production. One of the best fits between pasture production and animal requirements is seen in the fat-lamb industry of New Zealand (Hamilton 1956). The pasture reaches peak production in spring and early summer. This is followed by a low-production period, then a smaller autumn peak, followed by the dormant winter period. Cull ewes purchased in late summer are mated to lamb in early spring. Most of the lambs are sold off at 16–20 weeks of age when pasture growth declines in the summer. The remaining lambs are given the best of the pasture while the ewes are turned on to bare pasture to bring them down in condition. Prior to tupping they are run on better-quality autumn pasture and then returned to bare maintenance during mid-pregnancy. At this period grass is saved for feeding off to the ewes before lambing. At the peak periods of grass production cattle are often introduced in order to keep the grass down to the right stage of growth for the sheep. When the grass production declines the cattle are sold off fat. The live-weight gain from this system of farming ranges from 448 to 673 kg per hectare per year.

The New Zealand climate is very favourable for grassland farming. The animals are grazed out of doors all the year round. There is minimum conservation of hay or silage and very little feeding of concentrates. Some districts in New Zealand can produce under year-round grazing 15 700 kg dry matter per hectare having crude protein content between 20 and 30 per cent and with a starch equivalent of 8400 kg per hectare per year.

Conditions for grassland farming in the tropics are far less favourable. The dry season or season of minimum grass growth often extends to five or six months of the year and, during the latter part of this period, the quality of the pastures is not high enough to meet even maintenance requirements, much less the requirements of live-weight gain or milk production. This periodic nutritional deficiency retards the age of maturity. Heat stress and the prevalence of diseases, especially trypanosomiasis, aggravate the situation.

Beef-cattle production in the tropics is almost entirely extensive or free range, intensive production on improved pastures being restricted mainly to the experimental stations and areas farmed by European settlers. The land is generally communally owned and the nomadic pastoralists move with the rainfall and the pasture. Mixed farming is encouraged but seldom practised, the usual contribution of the pastoralist to the peasant farmer being the return of cattle droppings on land from which crop residues have been grazed.

The animal best suited to tropical nomadic conditions is the long-legged beast capable of covering considerable distances in search of forage and water and able to withstand long periods of semi-starvation. Among cattle breeds the slow-maturing Zebu or Zebu-cross meets these requirements.

Beef production in Nigeria

The following account (McIlroy 1962) is considered to be fairly typical of this industry in tropical West Africa.

Most of the Zebu cattle are owned by nomadic pastoralists, the Cattle Fulani and the Shuwa Arabs. Their seasonal migration is determined by the distribution of the tsetse fly and the quality of available pasture. Only the semi-arid portion of Northern Nigeria and the highlands of the Cameroons and the Bauchi Plateau are perennially free from tsetse but during the long dry season, October to April, a much wider area is free. The pasture in the semi-arid region provides sufficient nutrients for no more than three to five months of the year and falls below maintenance level in the dry season, when cattle are moved southward in search of water and better grazing. At the approach of the rains there is a movement in the reverse direction. The cattle progressively lose weight during the latter part of the dry season and show consistent live-weight gains only in the spring and summer. The weight lost is recovered in four to six weeks in the subsequent rainy season, but the set-back experienced is one of the most serious factors restricting production, as it retards maturity and increases the susceptibility to disease.

Experiments at Shika stock farm in the northern Guinea savanna zone have demonstrated that under conditions of more adequate feeding of the growing animal, Zebu bulls may be brought to 252 kg and heifers to 234 kg live-weight at eighteen months instead of three years. It has been shown by Miller (1960) at Samaru in the same zone that the feeding of high-protein concentrates (*Sorghum* grain and blood meal) to Zebu heifers, grazing low-quality predominantly *Hyparrhenia* pasture during the latter part of the dry season and early rains, leads to a gain of 0·23 kg per head per day in live-weight during the dry season.

To ensure continued productivity of the natural grasslands of the northern Guinea savanna, long rest periods in conjunction with regular burning at the beginning of the rains appear to be necessary.

Range grazing in the derived savanna of the Western Region

Figures obtained at Fashola (Ogor and Hedrick 1960) indicate that about four times as much land is required for bush or rough grazing as is required for sown pasture. Experience has shown that while 2·4 hectares of bush grazing is adequate during the wet season, 7 hectares is necessary in the dry season (Table 8).

TABLE 8

The development and carrying capacity of pastures in the derived savanna of the Western Region of Nigeria (Ogor 1961)

	Suggested area of forage crops required per animal unit		
		Average per animal unit	
Type of pasture	Period of year	Seasonal	Total
Bush grazing	Growing season	2·4	
	Dry season	7·2	9·6
Improved pasture	Growing season	0·3–0·4	
	Dry season	2·0	2·4
Plus silage for dry season	Maize	0·1	
	Grass	0·2–0·26	

Note. 1 animal unit = mature Ndama cow, 320 kg, 3½–4 years old.

The high requirement in the dry season makes it desirable to provide small areas of improved pasture to help carry livestock through this period. It has been shown by Ahlgren (1959) and Hedrick (1960) in trials at Agege (rain forest) and Ilorin (derived savanna) that elephant grass (*Pennisetum purpureum*) and *Andropogon gayanus* are productive grasses for the drought period in association with the legume *Stylosanthes guyanensis*.

At Matapos in Rhodesia under a 40 cm summer rainfall the ratio of natural grassland to supplementary fodder crops is 12:1, or 240 hectares of grazing land associated with 20 hectares of fodder crops.

In Mysore State in India, under a monsoon rainfall, the proportions are 80 hectares grazing land, 40 hectares natural hay land, and 20 hectares under an ordinary arable rotation, incorporating fodder crops (Whyte 1960).

In any system of bush grazing, in order to prevent local overgrazing, the distribution of the animals over the pasture area must be controlled by fencing or herding, or rotational grazing. Water points should not be more than three miles apart.

Animal health on ranges

The following routine inoculations against prevailing diseases are given to all stock on range: lapinized rinderpest vaccine, pleuropneumonia vaccine, anthrax spore vaccine, and haemorrhagic septicaemia vaccine. It is observed that helminth parasites build up to dangerous levels in the humid tropics. A system of rotational grazing with rest periods of at least six to eight weeks is desirable.

Live-weight gains

Daily gains of young Ndama* heifers grazing range pastures of *Cynodon plectostachyus* and *Panicum maximum* at Fashola during the rainy season (April–June) ranged from 0·9 to 1·4 kg live-weight.

* The Ndama are small beef cattle imported from Guinea. They are to some extent resistant to trypanosomiasis. The Ndama have all the characteristics of temperate-region beef cattle; they mature early and are reasonably fertile. They are most suitable for grazing improved pasture, either range or intensive. The mature weight rarely exceeds 340 kg.

Daily gains over the period July–October were: *Cynodon plectostachyus*, 0·22 kg, *Panicum maximum*, 0·32 kg, and, on a mixture of the two grass species, 0·40 kg live-weight. Over twelve consecutive months sixty Ndama heifers on *Cynodon plectostachyus* at Fashola averaged 0·21 kg per day live-weight gain with live-weight production per hectare per year of 269 kg (Ogor and Hedrick 1960).

In preliminary trials White Fulani (Zebu) heifers on *Cynodon plectostachyus/Centrosema* pasture at University College, Ibadan (Oyenuga 1958), in May–June averaged 0·57 kg daily live-weight gain; in the semi-dry season, October-December, 0·43 kg, and, in the dry season December–March, barely maintained weight. The stocking rate was 4 beasts per hectare. In a second trial with young Zebu stock, two animals at a stocking rate of 4 beasts per hectare, grazing *Cynodon plectostachyus*, made average daily live-weight gains of 0·31 kg over a period of 411 days (October 1956–December 1957), and three animals, at a stocking rate of one beast to 0·36 hectare on *Cynodon plectostachyus/Centrosema*, gained 0·36 kg live-weight per day over 177 days in December 1957–May 1958 (Hill, unpublished). The live-weight gain per hectare per year on the *Cynodon plectostachyus* was thus 461 kg. In replicated trials on the University farm, Ibadan, in 1960–1, fifteen two-year-old Ndama steers grazing *Cynodon plectostachyus/Centrosema* pasture rotationally, with rest periods of eleven weeks and no additional supplementary feed over twelve consecutive months, averaged 0·20 kg live-weight increase per day. The stocking rate was one beast to 0·24 hectare and the production per hectare per year 307 kg live-weight. Weight was lost towards the end of the dry season when the grass became coarse and fibrous. During the rains the dry matter of the pasture contained 16 per cent crude protein and 32 per cent crude fibre; towards the end of the dry season the crude protein content fell to 8 per cent and the crude fibre increased to 35 per cent. It is apparent that sown pastures in the tropics are able to maintain live-weight production comparable to that which is realized on good permanent pastures in temperate regions.

In the Caribbean, remarkably high production figures have been recorded on pangola-grass pastures, an annual gain per hectare of 1178 kg live-weight being achieved on pangola grass fertilized with

335 kg of ammonium sulphate per hectare and 365 kg on unfertilized pangola grass (Oakes 1960).

The problem common to all tropical areas is how to maintain the health and live-weight of the stock during the dry months of the year. In the drier parts of Australia high-protein concentrates of linseed meal, urea, or lucerne chaff have been used successfully to supplement the inferior pasture in the dry season. Concentrates are expensive. Less costly alternative supplements are a legume crop grown with or without irrigation for feeding as soilage, the conservation of saved pasture as hay and silage, and the reservation of paddocks of *Sorghums*, e.g. Sudan grass, for feeding out during the dry season. Browse trees and shrubs are of considerable value in providing a limited amount of fodder of high feeding value in some areas, e.g. the Sudan, where the leaves and pods of *Acacia* species in particular provide the equivalent of a protein concentrate although they are high in fibre.

12 Grassland Improvement

THE renovation or improvement of tropical grasslands must be approached from two directions; the management of the grazing animal must be considered in relation to the improved management of the pasture.

PASTURE MANAGEMENT

Re-seeding

Deteriorated or derelict permanent pastures in temperate regions are normally renovated by ploughing and re-seeding with superior species of grasses and legumes. This procedure, despite serious criticisms, has become so general that it has been said that the only pastures that should not be put under the plough for renovation are those that are unploughable for reasons of inaccessibility, e.g. hill pastures. Voisin (1960) has pointed out that the enormous increase in production following ploughing and re-seeding of grassland is not maintained indefinitely and falls in the fourth year, and that the final production of the ploughed-up pasture becomes less than that of the unploughed pasture. He presents considerable evidence in support of this conclusion. Nevertheless, the practice continues.

In the tropics, one of the quickest methods of pasture improvement is replacement of low-producing grasses by better species and varieties of grasses and legumes. Caution should be exercised in the use of the plough, because of the risk of rain and wind erosion. Less drastic methods for the preparation of the seed-bed are the ploughing of widely spaced furrows in which the seed is sown, or use of the sod-seeder which re-seeds the existing sward directly. The aerial broadcasting of legume (clover) seed and superphosphate has proved successful in the renovation of otherwise inaccessible hill pastures in New Zealand and should be advantageous on tropical range lands.

A possible legume genus is *Stylosanthes*. A degree of success has been achieved in the introduction of legumes into tropical grasslands by drilling these in widely spaced rows across the sward. In the case of rhizomatous or stoloniferous grasses that have become sod-bound, considerable improvement may be effected by discing and oversowing with legume seed. The grass regenerates via the cuttings of rhizomes or stolons that remain in the soil.

Fertilizing

Restoration of soil fertility by the application of fertilizers, especially nitrogen and phosphate, leads to increased production on poor soils, e.g. on the poor soils of the savannas of Guyana, the application of fertilizers combined with the introduction of better grass species raised the carrying capacity fortyfold. In most tropical areas the present low demand for animal products renders the application of even minimal quantities of fertilizer uneconomic and legumes are the only practicable source of nitrogen.

Control of invading vegetation

On cultivated pastures, invading vegetation may be controlled by mowing and by selective herbicides. On range lands biological control has been used successfully, e.g. the prickly pear (*Opuntia* spp.) controlled in Queensland by caterpillars of the moth *Cactoblastis cactorum*.

Control of pests and diseases

Locusts, grasshoppers, and crickets all eat large quantities of tropical grasslands. Small-scale invasions are controlled by insecticidal sprays, while major invasions come within the aegis of international organizations.

Burning

Controlled burning at appropriate seasons is an effective means of grassland improvement. Burning late in the dry season removes old and dry vegetation of low feeding value and facilitates the growth of new grass in the early rains. The least damage to tree growth is caused if the burning is carried out soon after the first rains, but, if the grass and not the trees is the first consideration, the burning should be done late in the dry season.

Water development

When soil moisture is a limiting factor in grassland development the construction of dams, earth tanks, and reservoirs may lead to local improvement.

Supplementary feed

The provision of supplementary feed in the form of hay, silage, fodder crops, or 'standing hay' is one method of easing the grazing pressure on pasture during the dry season.

Tree growth

Provision of some shade is desirable, for example, it has been shown that the yield of *Axonopus compressus* (carpet grass) is 20 per cent higher under leguminous shade, and the protein content is also higher. Encroachment of bush must, however, be avoided as this reduces the carrying capacity of the pasture.

LIVESTOCK MANAGEMENT

Overgrazing and overstocking of grasslands lead to their deterioration, to soil exhaustion, and to erosion. Control must be exercised over the number and kind of livestock and their distribution in relation to the availability of the forage. On range lands the provision and location of water supplies, salt, and mineral licks assists in the proper utilization of the entire grazing area. Herding and fencing serve to exclude stock from resting pasture and control grazing. Rotational grazing has been shown to be of little or no advantage on pastures of low productivity so long as moderate stocking rates are employed. The following proposals have been made for the improvement of grasslands in tropical West Africa (McIlroy 1962).

(a) A limiting factor to pasture improvement at the present time is the lack of adequate sources of seed and planting material for the establishment of new grassland swards or the rehabilitation of deteriorated swards. (Author's note: This requirement has been met in East Africa and in Rhodesia where improved varieties of grasses are available in commercial quantities.) Even though the semi-natural indigenous swards will constitute the principal grazing areas these will need supplementation by smaller areas of improved pasture for dry-season feed and in the forest zone it is likely that sown pastures will be extended.

Consideration should be given to the establishment of a Central Grassland Research Station for the quarantine and evaluation of introduced species, with adequate nurseries, multiplication areas, and laboratory and glasshouse facilities. In each of the regions testing and seed-multiplication centres should also be established. The distribution of the plant material could well be effected through the regional agricultural extension services.

The Central Station would make observations of survival rate, vigour, growth habit, time of flowering and extent of seed production, seasonal productivity, response to cutting, resistance to drought and disease, and nutritive value of the grass and legume species. While it is unlikely that the introduction of further new species will be necessary, more productive or otherwise desirable varieties may well be found. There is room for selection and breeding within the existing strains of indigenous and introduced species.

(b) Chemical methods for the control of weeds and regenerating bush in pasture swards require study. Aerial spraying of weedkillers may be a possibility and may be useful in the establishment of pastures.

(c) Since nitrogen is a universal requirement of grass species and cannot be met by the manure bag, consideration should be given to the aerial broadcasting of *Stylosanthes guyanensis* (legume) seed. As the seed is extremely fine and light it would need to be pelleted with soil. Superphosphate could be broadcast by air at the same time over the savanna area of the Northern Region. An appropriate time for aerial broadcasting would be after the annual burn. The aerial application of legume seed and superphosphate has made possible the improvement and more efficient utilization of many thousands of acres of otherwise wasteland in New Zealand.

Foster (1961) has obtained encouraging results in the establishment of *Stylosanthes* in the range—and pastures of Northern Nigeria by the agency of grazing animals. *Stylosanthes* seed mixed with *Sorghum* meal was fed to cattle, which then distributed the seed over the range land in their dung.

SOME EXAMPLES OF GRASSLAND IMPROVEMENT PRACTICES

India

Systems of rotational and deferred grazing that provide correctly timed rest periods and opportunities for natural re-seeding are being introduced at the village level. The aim is to reduce the present stocking rate of 7·5 animals per hectare during the entire year to 1·25 animals per hectare for four months of the year and to replace grazing with hand-feeding during the remaining months. Land is set aside for the cultivation of fodder crops, and residues from arable cropping are also used for supplementary feeding.

Irrigated pastures have been established on government and military farms with introduced grasses including *Pennisetum purpureum*, *Panicum maximum*, *Brachiaria mutica*, and *Sorghum sudanense* (Whyte and Kumar 1957).

Measures for the renovation of natural grasslands involve controlled grazing, manuring, and re-seeding, and erosion control by contour furrowing. Regulation of burning is also being attempted (Dabadghao 1960).

More recently improvement has been effected by the introduction of legumes *Atylosia* and Siratro into *Heteropogon* communities. These communities have also been re-seeded with *Cenchrus ciliaris*, *C. setigerus*, and *C. fulrus* with consequent increase in protein content (Dabadghao and Sharkarnarayan 1970).

Colombia

About 90 per cent of the grasslands of Colombia are unimproved. In some instances 5 hectares are required to support one animal unit.

Experimental results have shown that improvement is best effected by destroying the old sod, planting and harvesting several cultivated crops, and seeding an improved grass alone or in admixture with a legume.

In the cooler climatic region good results have been obtained without destruction of the old sod by scarification with rotavator, disc harrow, or spike cultivator and overseeding with temperate-region grasses and legumes. A dressing of complete fertilizer assisted early growth.

At lower elevations in hot climate, pangola-grass stolons were transplanted successfully in furrows opened at 90 cm intervals in an old *Axonopus* pasture. The introduction of pangola grass into the pasture more than doubled the yield. Fertilizer application was not necessary for establishment of the cuttings (Rockefeller Foundation 1960–1).

Brazil

Natural grasslands constitute the majority of the land of southern Brazil.

Brush control together with phosphate application at Säo Gabriel increased the daily live-weight gains of grazing cattle 33 per cent, maintenance units 36 per cent, feed nutrients 45 per cent, and gains per hectare 96 per cent. At Vacaria it was found that higher stocking rates could be used in the warm season provided that areas were saved for the making of hay for winter feeding. By this system the annual return of beef per hectare was increased by 69 per cent compared with that from continuously grazed pastures. A combination of permanent native pasture with winter annuals and some hay feeding when neither of these pastures is in production is recommended (Grossman 1956).

West Indies

Grassland management trials conducted in Jamaica (1948–54), on indigenous pastures with *Stenotaphrum secundatum* and *Paspalum notatum* dominant, showed that production could be increased considerably by rotational grazing, mowing twice annually, and the application of fertilizers. Fertilizer application was, however, uneconomic, owing to the low market value of meat. Annual beef production per hectare ranged from 67·2 kg on control plots to 130 kg on plots that were mowed and fertilized. The carrying capacity of these pastures ranged from 0·5 to 1·1 hectares to the growing steer.

Pennisetum purpureum persisted for six years under grazing and carried 5 yearling steers per hectare in May to July and 4 steers per hectare for the rest of the grazing period. Supplementary feeding was necessary from January to March.

Trials with pangola grass (*Digitaria decumbens*) and coastal Bermuda grass (*Cynodon dactylon*) showed the carrying capacity of both grasses to be about 0·24 hectare to the steer without fertilizer. Fertilized pangola pasture carried one steer to 0·2 hectare (Motta 1956).

Pangola grass has made possible the rehabilitation of areas that had been otherwise ruined by hundreds of years of intensive cropping. It has proved to be one of the most important pasture grasses in the West Indies. While rotational grazing of pangola grass is essential for maintaining quality, rest periods should not exceed five weeks or the grass becomes too fibrous. In Trinidad the standard management procedure involves topping or light grazing six to eight weeks after planting, followed by use of a brush cutter. Finally a chemical weedkiller, e.g. 2,4-D, is applied at the rate of 3 kg per hectare (Forster, Wilson, and Butterworth 1960).

In the West Indies, pangola grass forms a dense ground cover which suppresses associated prostrate herbaceous legumes. The tendency is, therefore, to apply artificial fertilizer rather than to depend upon the legume as a source of nitrogen. In field trials the application of nitrogen at rates up to 270 kg of nitrogen per hectare at planting gave almost linear increases in forage yield.

Congo

Improvement of natural pastures in the lower Congo has been effected by the introduction of legumes of which *Stylosanthes guyanensis* proved to be the best species. *Stylosanthes* was sown at a rate of about 3 kg per hectare after the pasture had been burned. The improved pastures were protected from grazing for eight to nine months in the valleys and for about twelve months in the hills. A two-year old sown pasture of *Stylosanthes guyanensis* in admixture with *Brachiaria ruziziensis* carried about 785 kg per hectare in the dry season (Delhaye 1959).

At Yangambi, *Setaria sphacelata* and *Brachiaria ruziziensis* are the principal grasses of sown pastures, while at Gimbi good pastures were established by sowing of *Setaria sphacelata* in admixture with

Brachiaria mutica (INEAC, 1958, 1959, 1960). At Mulungu (Kivu) the natural pasture has been improved rapidly and cheaply by introduction of *Setaria splendida*, which grows well among the indigenous grasses (Compère 1961).

Tropical Australia

Grassland improvement methods developed at the Katherine Research Station in the Northern Territory of Australia are summarized in the progress report of that institution for 1954–6. The station is located about 200 miles inland from Darwin at latitude 14° S. The annual rainfall is 90 cm falling in December–March. The dominant vegetation is open woodland with tall grass.

Under the extensive management systems normally practised the carrying capacity is about two to three cattle per square kilometre. Deferred grazing systems—two weeks on and eight weeks off—have been practised for three years with satisfactory results on a pasture of *Themeda australis*, *Sorghum plumosum*, and *Chrysopogon fallax*. The legume *Stylosanthes humilis* has been successfully established in the local pasture. As it is unpalatable it spreads at the expense of the grasses when heavily grazed but when not cut or grazed is suppressed by the grasses.

Many grasses and legumes have been introduced in order to bridge the dry-season gap of up to six months of the year when cattle lose weight. *Cenchrus ciliaris* and *C. setigerus*, which have shown promise in the coastal areas of West Australia, have proved to be the most successful of the introduced grass species, and *Cenchrus/Stylosanthes* mixtures are giving promising results. *Stylosanthes humilis* in association with *Cenchrus* can yield 5 tons dry matter per hectare per year or with phosphate fertilizer 17·5–20 tons per hectare. Cattle maintain or gain weight in contrast to the loss of weight in the dry season on native pasture.

Shaw (1961) at Rodds' Bay in central Queensland showed that inclusion of *S. humilis* in native spear-grass pasture and annual application of 125 kg per hectare of molybdenized superphosphate increased the carrying capacity of steers by three times. The annual live-weights gain per hectare increased at least five times, and animals could be marketed one to two years earlier than on native pasture.

13 The Nutritive Value of Tropical Pasture Grasses and Legumes

Nutritive value and digestibility

WHILE the chemical composition of a herbage species is of value as a guide to its nutritive value in comparison with other feeds, feeding trials should be carried out with the animals for whom the herbage is intended and the results of these trials related to the results of chemical analysis.

Minson and Brown (1959) in trials at the Grassland Research Institute at Hurley in England have found that herbages with the same nitrogen content may have widely differing digestibilities. Even when corrections were made for species and for season of growth the relationship between herbage nitrogen content and herbage digestibility still had predictable errors greater than three digestibility units making them of little value for establishing differences in digestibility between different feeds. There is some evidence that the soluble carbohydrate content may be as important as the nitrogen content in determining the digestibility of a herbage by ruminants and that high nutritive value may be dependent upon the attainment of a proper balance between these two constituents.

The nutritive value of a herbage species is influenced by the leaf/stem ratio, by the stage of growth at cutting or grazing, by the soil fertility and manurial treatment, and by climatic conditions. Legumes are generally richer than grasses in nitrogen and in phosphorus and calcium.

An average sample of good leafy herbage of temperate origin and of about three weeks' growth might be expected to contain about 20 per cent of dry matter of the following chemical composition:

crude protein (N × 6.25)	22 per cent	(range 16–28);
ether extract	6 „ „	(„ 5– 7);
nitrogen free extract	47 „ „	(„ 43–49);
crude fibre	17 „ „	(„ 14–19);
ash	8 „ „	(„ 6– 9).

Tropical grasses are low in crude protein and high in crude fibre when compared with temperate grasses cut at similar stages of growth. The dry-matter yield of tropical species, on the other hand, often exceeds very considerably that of temperate grasses. The dry matter of Rhodes grass (*Chloris gayana*), cut four weeks after first cutting back, contained only 14·9 per cent crude protein and molasses grass (*Melinis minutiflora*) 9·8 per cent. Of a number of grasses analysed by Oyenuga (1957) at University College, Ibadan, Nigeria (Table 9), these species contained the highest and lowest amounts of protein respectively. No fertilizer was applied.

TABLE 9

Analysis of grasses from Nigeria

Herbage species	Crude protein % 3–4 weeks	Average	Crude fibre % 4 weeks	Average
Andropogon spp.	11·0	7·6	26·9	31·0
Chloris gayana	14·9	8·4	27·4	30·1
Cynodon plectostachyus	11·0	7·5	29·5	30·6
Melinis minutiflora	9·8	6·8	32·1	33·7
Panicum maximum	13·5	8·2	28·3	33·8
Paspalum conjugatum	10·7	6·6	29·5	30·2
Pennisetum spp.	14·0	9·2	26·0	30·9
Setaria sphacelata	10·9	6·5	30·8	33·0
Centrosema pubescens (legume)	15·8	16·9	30·3	30·7

Duckworth (1949) found that elephant grass (*Pennisetum purpureum*) and other coarse tropical grasses such as Guatemala grass (*Tripsacum laxum*), bamboo grass (*Panicum fasciculatum*), and Uba cane (*Saccharum sinense*) grown in the West Indies contained less than half as much digestible crude protein and approximately three-quarters as much starch equivalent as the finer grasses of temperate-zone pastures.

The relation of nutritive value to height of cutting of some grasses in Tanzania is illustrated in Table 10 (after French 1941).

Other things being equal, the leafier grasses are preferable for grazing as the leaf contains more protein and less fibre than the stem.

TABLE 10

Composition and nutritive value
(dry matter)

Species	Crude protein	Crude fibre	Digestible protein	Starch equivalent	Stage of maturity
Cynodon plectostachyus					
Short (20–30 cm)	20·25	23·97	15·32	49·14	No flower heads
Medium (30–45 cm) (second growth)	11·64	32·15	8·09	44·60	Few „ „
Long (60 cm)	11·53	31·72	8·14	34·96	All stages of maturity
Digitaria pentzii					
Short (10–15 cm)	14·94	30·16	11·70	54·75	No flower heads
Short (15–23 cm)	9·24	33·59	5·30	45·73	Few „ „
Short (25–30 cm) (second cut)	7·28	37·97	3·71	37·40	Anthers emerged
Panicum maximum					
Medium (30–45 cm)	9·15	31·19	5·70	42·81	No flower heads
Long (75–90 cm)	9·31	34·46	3·83	35·42	„ „ „
Tall (leaves 120 cm stem 180 cm)	5·61	41·76	2·74	29·21	Early flowering
Pennisetum purpureum					
Medium (30–45 cm)	10·39	31·44	5·99	44·60	No flower heads
Long (75–90 cm)	9·54	30·37	4·86	35·17	„ „ „
Tall (up to 240 cm)	7·75	46·88	3·55	27·66	Anthers emerged
Cenchrus ciliaris					
Medium (30–45 cm)	11·64	33·93	8·18	42·44	Anthers emerged
Long (75–90 cm)	11·65	36·35	8·23	49·19	Early flowering
Chloris gayana					
Short (30–45 cm)	10·06	34·36	6·56	50·49	Few flower heads
Short (30–45 cm) (second growth)	7·50	41·03	4·45	38·20	„ „ „
Paspalum dilatatum					
Medium (leaves 30 cm)	7·19	34·17	3·31	46·69	„ „ „
Medium (second growth)	6·63	34·16	3·33	48·54	„ „ „

Note: Of the grasses listed *Cynodon plectostachyus* is the superior species.

The leaf/stem ratio of certain tropical grasses examined by Schofield (1945) is reported in Table 11.

TABLE 11

Proportion of leaf to stem of grasses cut in the mature stage

Grass	Leaf/stem ratio	Protein in leaf	Protein in stem
Brachiaria brizantha	1:2	6·8	4·0
Chloris gayana	1:2·4	8·1	4·2
Cynodon plectostachyus	1:2	10·8	4·6
Panicum maximum	1:1·1	7·7	5·3

In Trinidad, Paterson (1933) has shown that as the time interval between cuttings is increased, the protein content of elephant grass (*Pennisetum purpureum*), Guinea grass (*Panicum maximum*), and Para grass (*Brachiaria mutica*) falls while the dry matter increases. An interval of 8–10 weeks between cuttings generally gives the highest production of digestible nutrients. The interval may have to be extended during the dry season. Rains and Foster (1956–7) in Nigeria examined the yield and composition of Gamba grass (*Andropogon gayanus*) cut at heights of 60, 90, 120, and 150 cm during three successive years.

The total number of cuts at the different heights was 12, 9, 8, and 7 respectively. The dry-matter yield was greatest for the plots cut at 150 cm. The mean crude protein yield was highest from those cut at 120 cm while the plots cut at 60 cm had the highest crude protein content. Nordfelt *et al.* (1951) in studies of Napier grass in Hawaii found that while the yield of dry matter was highest for a fourteen-week cutting interval, the yield of crude protein was highest at a ten-week interval, and the crude protein content was highest when the cutting interval was six weeks.

In Nigeria, Oyenuga (1960*a*) found that Guinea grass (*Panicum maximum*) gave the highest growth rate and green-fodder yield (70 tons per hectare per year) when allowed a twelve-week growth cycle. Cutting every three weeks gave a total minimum yield of 50 tons per hectare. The highest yield of dry matter was also obtained for the twelve-weekly cut. The percentage crude protein in the dry matter for three-weekly cuts was about 11·6 and declined to 6·4 per cent when the grass was cut every twelve weeks. The effect of frequency of cutting on the yield and chemical composition of *Panicum maximum*, *Tripsacum laxum*, and *Andropogon tectorum* has been reported by Oyenuga (1959, 1960*b*).

The effect of stage of growth on the crude protein and crude fibre content of tropical grasses in the Sudan is illustrated by the analyses in Table 12 made by Boyns (1955) at the School of Agriculture of the

TABLE 12

Composition of grassland species in the Upper Nile Province, Sudan (% of dry matter)

Species	Age of growth in weeks								
	2	4	6	8	10	12	14	16	Mature
Chloris gayana									
Crude protein	8·56	7·81	7·87	4·88	4·53	4·87	3·60	3·52	3·58
Crude fibre	30·51	35·38	35·91	35·74	35·93	38·19	35·16	35·49	35·70
Cynodon dactylon									
Crude protein	7·87	9·01	10·62	7·87	6·96	7·05	6·30	6·21	5·70
Crude fibre	37·64	36·16	35·04	40·83	39·09	43·44	41·63	32·70	31·71
Echinochloa pyramidalis									
Crude protein	6·43	8·05	6·27	5·36	3·90	4·10	6·20	3·23	4·42
Crude fibre	34·92	37·72	39·45	40·18	39·81	39·68	34·54	37·47	40·26
Hyparrhenia rufa									
Crude protein	7·76	5·55	6·05	4·98	3·62	3·73	3·30	3·66	3·19
Crude fibre	31·78	33·23	34·93	37·61	41·80	44·76	34·74	35·76	40·25
Ischaemum brachyetherum									
Crude protein	8·70	7·03	6·43	4·93	3·90	4·00	—	—	2·80
Crude fibre	41·78	43·64	42·91	42·34	43·83	42·35	—	—	41·03
Panicum repens									
Crude protein	6·06	7·48	6·52	4·49	5·57	5·43	5·60	4·27	3·31
Crude fibre	42·91	36·65	38·66	40·13	41·85	38·35	42·26	42·75	39·18
Pennisetum ramosum									
Crude protein	8·40	6·09	5·83	4·52	5·90	5·31	—	—	5·20
Crude fibre	35·89	38·14	43·13	40·23	35·60	38·47	—	—	40·85
Setaria incrassata									
Crude protein	9·30	8·04	7·90	6·86	—	—	—	—	1·62
Crude fibre	36·82	32·50	34·69	33·19	—	—	—	—	40·55

Note: Annual rainfall 800 mm. Rainy season: May–October.

University of Khartoum, Sudan, and quoted in *Grasses in agriculture* (F.A.O. Agricultural Study no. 42, p. 146).

The seasonal fluctuation in the composition of veld (natural pasture in Salisbury, Rhodesia) as determined by Weinmann (1955), is recorded in Table 13 (figures in percentages of dry matter).

TABLE 13

Seasonal fluctuation in the composition of veld

	Jan.	Feb.	March	April	May	June	July
Crude protein	7·40	6·33	4·73	3·73	1·95	1·89	1·92
Crude fibre	32·23	38·00	43·61	43·67	42·71	44·67	44·31
P_2O_5	0·43	0·49	0·33	0·24	0·30	0·15	0·16
CaO	0·59	0·48	0·42	0·34	0·42	0·39	0·39

The influence of stage of growth on the mineral and vitamin contents of tropical grasses has not been reported but it has been shown that, in temperate regions, the younger and leafier the herbage the greater the content of calcium, phosphorus and potassium. Waite and Sastry (1949) have established a close correlation between the protein content of a grass and its carotene content. Vitamin C is similarly related to the age of the plant. The younger the grass the richer it is in vitamins.

The influence of nitrogen fertilizers on the yield and composition of tropical grasses

Large increases in yield and protein content are normally obtained by the application of nitrogenous fertilizers to grasses. Table 14, after Ripperton and Takahashi (1948), shows a maximum increase as the result of the application of 336 kg ammonium sulphate per hectare after each grazing. The number of grazing days was increased from 204 to 411 per year at this level of nitrogen.

Studies of nitrogen fertilization of pastures at Marandallas Grasslands Research Station in Rhodesia (1957) have shown that greater yields are obtained by applying nitrogen in one dressing at the start of growth than by applying the same amount in several smaller

TABLE 14

Influence of nitrogen fertilizers on yield and composition of tropical grasses

Ammonium sulphate applied after each grazing (kg/ha)	Ammonium sulphate applied per year (kg/ha)	Dry matter per year (kg/ha)	Crude protein (%)
Nil	Nil	7157	7·00
168	773	12208	8·75
336	1719	17225	11·37
672	3864	15512	16·25

dressings over the growth period. A temporary increase in the protein and non-protein nitrogen (N.P.N.) content of the grasses occurred about one week after the date of application of nitrogen. On a *Setaria* pasture nitrogen fertilizers had little effect on the gain per animal but considerably increased the gain per hectare.

TABLE 15

Chemical composition of tropical legumes grown in Queensland (Walsh 1958) (% of dry matter)

Plant	Season	Crude protein	Lime	Phosphate
Centrosema	Wet	23·8	3·61	0·84
	Dry	15·8	1·44	0·49
Pueraria	Wet	19·5	2·63– 1·047	0·607– 0·42
Stylosanthes	Wet	18·05	2·70	0·75
	Dry	10·55	1·27	0·49
Calopogonium	Wet	19·5	2·17	0·584

14 Conservation

Hay

THE best time to harvest grass for hay is in the early flowering stage. When cut in the early growth stage nutritive value is higher but yield is low and the moisture content is high. When cut after flowering the increased yield does not compensate for the decreased nutritive value and palatability. Rapid drying with minimum exposure to sun and rain is essential to avoid excessive loss of nutrients. Further losses occur in handling and in storage. Careless handling leads to loss of leaf and storage under humid conditions to mould formation.

It has been estimated that during hay-making and storage the average loss of dry matter is about 25 per cent and, in a bad season, as much as 50–60 per cent of the nutritive value may be lost. Respiration losses after cutting amount to about 10 per cent; mechanical damage to 5–10 per cent and bacterial and enzyme action in a stack of imperfectly dried hay to about 5–10 per cent.

Losses in drying in the field may be reduced by the use of tripods, hurdles, and racks and, where power and mechanical equipment are available, partially dried hay may be carted to the barn for drying by heated air. These methods all add to the cost.

The high rainfall during the growing season in the humid tropics makes hay-making almost impossible. Weather conditions are much more favourable in the semi-arid tropics but the feeding value of the product is often little better than that of 'standing hay' or deferred grazing.

Artificially dried grass

Artificial drying of grass eliminates the losses of dry matter that occur in hay-making or silage manufacture. As the cut grass is carted in the moist state there is no loss of leaf and carotene loss is minimal. The process requires special equipment and is only economic when the grass is of very high quality and the product equal in feeding value to a protein concentrate. Few tropical grasses are of sufficiently high quality, even at the most nutritious stage, to justify the high cost of artificial drying.

Silage

Since silage manufacture is practically independent of weather conditions it is the most suitable method of conservation under tropical conditions. If well made, the product closely approaches in feeding value the grass or other forage crop from which it is prepared. Freshly cut herbage (or other suitable crop) is compressed in a stack, clamp, pit, or silo. Expensive concrete silos are not necessary, and excellent silage has been made in pits in West Africa, East Africa, the Sudan, and the West Indies. The pit should be dug on ground that is high and dry and adequately drained. Hillsides are ideal. The pit should be near the feeding area. The walls of the pit may with advantage be plastered with mud or cement. For collection of the material for silage manufacture a silorator or a tractor fitted with a buckrake is convenient. Compression in the pit may be effected by a tractor or heavy roller.

The nutritive value of the product is largely determined by the growth stage of the grass at cutting, while the quality depends on the type of fermentation. Adequate compression of the material restricts loss of carbohydrate by respiration, but sufficient air is left in the mass to allow aerobic bacteria to convert carbohydrate into acetic, propionic, and lactic acids. The acidity developed suppresses undesirable (butyric) fermentation. Butyric-acid organisms cease growth at about pH 4·2. Inadequate compression, as with stemmy material, leads to excessive respiration and overheating, with high losses of dry matter, reduction in the digestibility of protein, and loss of carotene. Overcompression, as with succulent herbage, leads to underheating, giving a foul-smelling unpalatable product. Good silage is made over the temperature range 10–38 °C. Chopping or crushing of stemmy or coarse material assists consolidation, and wilting is desirable before ensiling material of high moisture content. High carbohydrate and low protein content provide good conditions for successful silage manufacture.

For the ensiling of young grass and legumes the incorporation of additives that act as accelerators of fermentation, e.g. molasses or ground cereal grains, which promote the production of lactic acid, or as mild antiseptics, e.g. sodium bisulphite, is desirable. The most common additive is molasses, which is added to 3–4 per cent by weight diluted with equal parts of water and applied as a spray.

Breirem and Ulvesli (1960) give the following values for good quality grass silage.

pH	4·2
Lactic acid	1·5–2·5 per cent
Acetic acid	0·5–0·8 per cent
Butyric acid	<0·1 per cent

Ammonia nitrogen as percentage of total nitrogen not to exceed 5–8.

15 The Measurement of Grassland Productivity

THE measurement of grassland productivity has been reviewed recently in a series of papers presented at the 6th Easter School in Agriculture at the University of Nottingham (Ivins 1959). Methods of surveying and measuring vegetation form the subject of a bulletin by Miss D. Brown (1954). Both these works should be studied by those who contemplate entering the field of grassland research. The following outline of the methods of measuring grassland production is based on a review by Miss M. E. Castle (1955).

Methods based on botanical composition

Methods of botanical analysis and their application to pasture evaluation have been reviewed by Miss D. Brown (1954). The method commonly used is point-analysis (Levy and Madden 1933) using a point-quadrat. The apparatus consists of a rack with a series of holes through which long rods with fine points are thrust. Two types are in use, those in which the rods travel in a vertical plane and those in which the plane of travel is inclined at 45° to the vertical. As the points are thrust through the herbage a hit is recorded for each species. The analysis is based on the recording of every hit made by the point of each pin. It has been shown that vertical pins are less satisfactory than inclined pins.

For the inexperienced operator, manual separation and weighing of the species in a weighed sample is more efficient and less biased, though laborious and time-consuming. The herbage is clipped at standard height (usually as near ground level as possible) from a number of quadrats (the space enclosed by a small grid, often 15 × 15 cm) and the cut herbage is sorted into different species which are then weighed.

Methods of measuring the quantity of herbage

The clipping technique generally adopted consists of the cutting of a measured sample area of the pasture, weighing of the herbage, and expressing of the yield per unit of area. To overcome the problem of herbage growth during the period when the plots are being grazed, wire cages have been used to protect an area of the sward. The value of these cages is suspect since Cowlishaw (1951) has shown that, without grazing, yields over a fortnightly period are significantly greater under a cage than outside. The difference may be attributed, at least in part, to the different micro-climate within the herbage enclosed by cages and hurdles. When the plots are grazed for very short periods, e.g. two or three days, the herbage growth may be safely ignored.

In order to determine the quantity and quality of the herbage consumed by the grazing animal, i.e. the effective utilization, a measured portion of the pasture available to the animal is harvested by clipping with shears or an auto-scythe and the weights of dry matter, crude protein, etc. offered for grazing are then calculated. The unconsumed residues left on the plot after grazing are then determined similarly. The differences between the two sets of figures give estimates of the quantity and quality of the herbage consumed. The best results are obtained when the animals graze small areas for short periods but, even then, the error may be of the order of 20 per cent (MacLusky 1955).

Sources of error are listed by Sampford (1960) as

(a) variation in yield between plots;
(b) random sampling error in estimation of yield;
(c) variation in consumption resulting from differences of availability, palatability, and appetite;
(d) random sampling error in estimation of residue;
(e) errors in cutting, weighing, etc., resulting in biased estimates of yield and residue.

Despite the magnitude of the possible error, the herbage clipping technique is rapid and convenient and capable of yielding useful information when the sampling process is carefully standardized.

Methods of measurement in grazing days

This measurement is generally expressed as the product of the number of animals and the length of grazing time per unit area of grassland, and is useful in that the animal itself is the measure of pasture evaluation.

Methods of measurement in terms of animal produce

This involves expression of pasture output in terms of meat or live-weight increase per hectare. If milk yield is used, it should be remembered that this depends primarily on the stage of lactation of the grazing animal.

Methods of measuring the net contribution of grassland to the food requirements of stock using a standard food unit

This method consists of estimating the starch-equivalent, or total-digestible-nutrients (T.D.N.) requirements of the farm stock for their maintenance, growth, live-weight increase, milk production, and work. From this total is deducted the starch-equivalent value of all foodstuffs other than grass or grassland conservation products, and the difference is assumed to be a measure of the productivity of the grassland. The starch-equivalent method has the advantage of expressing results in a common unit, and takes into consideration both the maintenance and production requirements of the stock. The method and its application are fully described by Barker *et al.* (1955).

Methods based on the food-replacement technique

These methods are of little scientific value and are not discussed here.

Methods based on monetary returns

Biochemical methods

These methods are based on

(a) determination of the digestibility of cut herbage by digestibility trials on animals housed indoors in suitably constructed cages or stalls;

(b) collection and weighing of the faeces of animals grazing at pasture and calculation of the weight of herbage grazed (Raymond 1954).

They have recently been reviewed by Corbett and Greenhalgh (1959). If the quantity of organic matter (O.M.) excreted in faeces in a given period and the apparent digestibility of the organic matter of the herbage consumed are determined, then the quantity of organic matter consumed by the animal in that period can be obtained from the expression

$$\text{Herbage intake} = \frac{100 \times \text{weight of faeces O.M.}}{100 - \text{digestibilty of herbage O.M.}}.$$

The digestibility of the herbage organic matter is calculated from the expression

$$\%\,\text{digestible organic matter} = 100 \times \frac{\text{Feed O.M. eaten} - \text{faeces O.M.}}{\text{Feed dry matter eaten}}.$$

Corbett and Greenhalgh point out that organic matter is preferred to dry matter not only because herbage and faeces may be contaminated with soil but because, for herbage, digestible organic matter is practically identical with total digestible nutrients and, if adjusted for crude fibre (*Rations for Livestock*, Bull. **48**, 14th ed., HMSO, 1957), yields values roughly equivalent to those obtained from calculation of starch equivalent (Dijkstra 1954).

In order to avoid the inconvenience and expense of faeces-collection bags attached to grazing animals, the faeces output is now generally determined by the administration of inert reference substances (markers) which pass unchanged through the gut and are completely recoverable from the faeces, e.g. when the regular dose of marker in a given period is 10 g, a sample of faeces containing 5 g of organic matter and 0·1 g of marker yields an estimated output of 500 g of faeces organic matter during the period. Chromic oxide is the most promising marker. It is quantitatively excreted in faeces and requires fairly short preliminary and collection periods. It is usually administered in powder form in gelatin capsules or in the form of paper strips impregnated with the marker. Under grazing conditions, variations in level of food intake, feeding behaviour, and rate of passage of food through the gut lead to irregularities in marker excretion. Accordingly bulked samples of many defecations (30–50) collected from the sward should be analysed (Raymond and Minson 1955). Digestibility values determined by faeces collection from the sward differed little from values determined by the harnessed collection-bag method. Collection periods of five days with four animals should be adequate to reproduce mean faecal excretions within $\pm$ 5 per cent of actual, provided faecal chromic oxide is determined from truly representative samples.

The digestibility of grazed herbage can be estimated from the proportion of certain constituents in the faeces without the necessity of sampling the herbage. Faecal nitrogen appears to be the best internal indicator. It is easily determined by a standard procedure and gives a slightly more accurate prediction of the feed: faeces ratio than chromogen (Raymond 1948; Lancaster 1949). The weight of metabolic faecal nitrogen varies directly with the weight of food eaten and so the concentration of faeces total nitrogen rises with increasing digestibility. A number of regression equations relating faeces nitrogen concentration to herbage organic matter digestibility have been published (Minson and Raymond 1958), but the accuracy of the predicted digestibility depends upon the herbage used in the derivation of the equation, and a new regression equation should be derived for each pasture examined.

The errors of the faecal-nitrogen indicator method for estimation of herbage digestibility have been stressed by Raymond and co-workers at the Grassland Research Institute (Minson and Brown 1959; Raymond *et al.* 1959). Nevertheless, faecal-indicator techniques provide the most satisfactory estimates presently available of the amounts of herbage consumed by grazing animals. When the animals graze as a group it has proved possible to identify faeces sampled from the sward for nitrogen and chromic oxide by dosing each individual animal regularly with coloured polystyrene particles (Minson *et al.* 1960).

The following manuals are recommended to all grassland research workers:

Bulletin 45, Commonwealth Bureau of Pastures and Field Crops,

Research techniques in use at the Grassland Research Institute (Hurley 1961).

Bulletin 47, Commonwealth Bureau of Pastures and Field crops, *Some concepts and methods in sub-tropical pasture research* (1964).

G. R. Moule, *Field investigations with sheep, a manual of techniques* (C.S.I.R.O., Melbourne, 1965).

Joint Committee American Society of Agronomy, American Dairy Science Association, American Society for Animal Productivity, American Society for Range Management, *Pasture and range research techniques* (Ithaca, N.Y. 1962).

16 Pasture Quality

THE components of pasture quality are nutritive value, intake, and acceptability to the grazing animal. The best assessment of the quality of a pasture is made by the immediate consumer, which is the grazing animal. Output per animal is determined by nutritive value and intake and is thus synonomous with pasture quality.

Factors that control output per animal are illustrated in the following chart (after Barnes 1965):

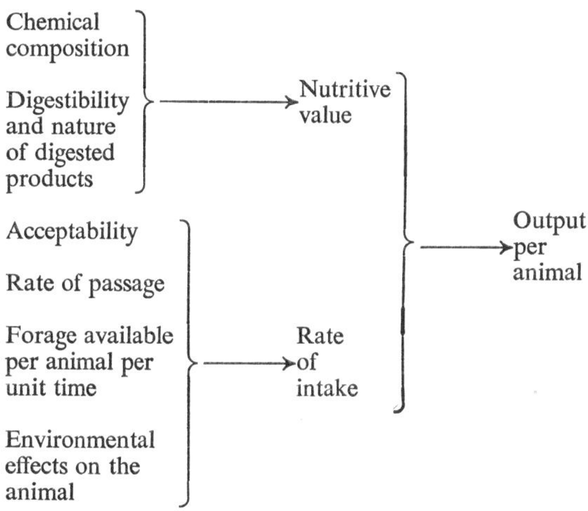

Minson (1968) suggests a three-stage pasture evaluation programme:

(a) The quality of a large number of pasture samples is measured by one of the indirect laboratory methods.

(b) Five or six of the most promising treatments are selected for more rigorous screening in which the intake of digestible energy is measured with pen-fed wethers.

(c) The final stage of quality evaluation involves the study of two or three of the best treatments in grazing trials with sheep, steers, or dairy cows at *low grazing pressure.*

After this evaluation, the best treatment should be tested in a full-scale grazing trial at several stocking rates in which pasture availability is also a factor determining animal production.

Digestibility

Recent reports from the Grassland Research Institute, Hurley (1966) have indicated that digestibility of the pasture is a useful index of quality and can be used effectively in pasture management. It was observed that if management maintained the pasture at a level of 70 per cent digestibility a daily production of 9 litres fat-corrected milk per animal-day for an extended period was achieved. When the digestibility dropped below 70 per cent, milk production declined also.

The digestibility, conventionally determined by feeding of the material cut from the sward to sheep held in pens, is now more conveniently determined by the two stage *in vitro* rumen fermentation technique developed by Tilley and Terry (1963) and subsequently modified (Rogers and Whitmore 1966; Barnes 1966). It has been shown by McLeod and Minson (1970) that the *in vivo* digestibility of mixtures of grasses and legumes as well as of single species can be estimated by the *in vitro* technique provided that suitable grass and legume standards are included. Cellulolytic enzyme digestion has recently been developed as an alternative to *in vitro* rumen fermentation. The indigestible residue is highly correlated with *in vivo* digestibility (Jarrige, Thevend, and Demorquilly 1970).

Digestibility estimates based on the evaluation of cut herbage ignore the selectivity of the grazing animal. Since grazing animals tend to select the more digestible components and leave ungrazed the overmature or soiled parts of a pasture sward, the digestibility based on feeding trials with cut herbage is likely to be underestimated. Weir and his co-workers (1959), using animals fitted with oesophageal fistulae, showed that sheep at pasture selected herbage with a higher crude protein and lower crude fibre content than was estimated in samples obtained by hand clipping. These results have since been confirmed with tropical pastures in Uganda (Bredon, Torell, and Marshall 1967). Such differences cannot be attributed entirely to selectivity, however. Lambourne (1964) found samples collected from fistulae in pen-fed animals to have higher dry-matter digestibilities than the original feed. He attributed the higher figures to the addition of soluble dry matter such as saliva. Similarly, the soluble carbohydrate content of fistula samples may be much lower than that of the feed. Despite these limitations, the oesophageal fistula technique allows collection from the grazing animal of the portion of the herbage actually consumed. Determination of the *in vitro* digestibility of this sample should give a fair approximation to the digestibility of the feed consumed.

The digestibility of grasses is considerably influenced by stage of growth. It reaches a peak just prior to ear emergence and then declines rapidly (Minson 1960). Nitrogen fertilization may reduce the digestibility of grass/legume swards when it depresses the proportion of legume, or it may increase the digestibility of herbage that has too low a nitrogen content to permit efficient operation of rumen microflora.

It has been generally assumed that the leaf/stem ratio influences digestibility, leafy varieties being more digestible than stemmy varieties. However, Dent (1963) and Minson and coworkers (1964) found certain stemmy varieties of grasses to be at least as digestible as leafy varieties. It is only when stems are allowed to mature that they lead to a marked depression in digestibility.

The digestibility of tropical grasses is in general about 13 units lower than that of temperate species. This difference is closely associated with climatic differences, digestibility being depressed by higher temperatures. The temperature effect is not observed, however, with tropical legumes (Minson and McLeod 1970).

Intake

The level of intake influences the digestibility. The immediate effect of increasing the intake is to increase the rate of passage through the gut and so to reduce the digestibility (Blaxter 1956). Increase in the stocking rate of animals at pasture reduces intake by reducing the availability of herbage, and leads to an overestimate of digestibility.

It has been generally assumed that animals will eat more of herbage species that are of high digestibility, and that low food intake is the result of the slower rate of digestion of less digestible foods and of the consequent greater quantity of residue that has to be moved through the digestive tract. The relationship between intake and digestibility is, however, variable. Hutton (1962) found only a slight decline in intake as the digestibility of herbage dry matter fell from 77 to 71 per cent. Minson and co-workers (1964) found that the intake of timothy grass (*Phleum pratense*) was less than that of other herbage species of similar digestibility.

Acceptability (palatability) has a marked effect on intake. Greenhalgh and Reid (1967) investigated the separate effects of palatability and digestibility on intake using rumen-fistulated sheep. The effects of palatability and digestibility appeared to be of approximately equal magnitude. Factors other than digestibility that may affect intake are rate of digestion, chemical factors, physical factors, e.g. hairiness, and external factors such as fungal infections or faecal contamination of the herbage. High levels of nitrogen fertilization can have an adverse effect on intake (Reid, Jung, and Murray 1966).

It is well known that the intake of legumes is higher than that of grasses (Reid and Jung 1965; Van Soest 1965). At 55 per cent digestibility, the intake of the tropical legume *Glycine wightii* was more than twice that of the tropical grass *Setaria sphacelata* (Milford and Minson 1965).

Among legumes, as among grasses, there is considerable species variation in intake. Osbourn, Thomson, and Terry (1966) found that the voluntary intake of sainfoin (*Onobrychis viciifolia*) of digestibility 62·6 per cent was 18 per cent higher than the intake of red clover (*Trifolium pratense*) of digestibility 64·6 per cent. Analysis of sainfoin and red clover showed little difference except for a higher lignin content in the sainfoin.

Ingalls (1965) concluded that 70 per cent of the variation in production potential (animal output) can be accounted for in terms of differences in voluntary intake, compared with 30 per cent by differences in digestibility.

Net energy value

It has long been recognized that the net energy value gives the nearest approach to the true value of a feeding-stuff for maintenance and production. The net energy is the gross energy less the energy lost in rumen fermentation, urine, and faeces. As the main source of energy for the ruminant is the volatile fatty acids (acetic, propionic, and butyric acids) produced by micro-organisms in the rumen from carbohydrates, determination of the amount and relative proportions of these acids and the efficiency of their utilization by the animal will give a measure of the nutritive value of the herbage.

Tilley and co-workers (1960) found that the ratio of acetic acid to propionic acid in the rumen liquor of sheep fed on different grass species ranged from 3·1:1 to 1·7:1. Acetic acid appears to be less efficiently used as a source of energy for fattening than are propionic or butyric acids. Acetic acid, on the other hand, is the precursor of milk fat while propionic acid is involved in the synthesis of milk solids (Rook 1959). For total milk production, acetic acid appears to be used more efficiently than propionic acid (Balch 1960). The individual fatty acids produced in the rumen thus serve different roles in animal production, and the type of grassland herbage that is suitable for one class of animal production may not be equally suitable for another class.

Energy-metabolism experiments (Blaxter 1960) indicate that there is a close relationship between herbage digestibility and net energy values, although as digestibility decreases net energy for animal production falls off more rapidly than digestibility.

Chemical composition and nutritive value

The crude protein content of herbage has been used as an indicator of nutritive value. The higher the protein content, the higher, in general, is the nutritive value.

Milford and Minson (1964) found that the intake of two subtropical herbage species declined significantly only when their crude protein content was 7 per cent or less. Glover and Dougall (1960) as a result of the statistical examination of published data found that the digestibility of total carbohydrate (crude fibre and

N.F.E.) began to fall markedly at crude protein levels below about 6 per cent. As the protein content of many tropical pastures is below this level for many months of the year, the intake of these feeds at these times is insufficient to meet the animal's requirements. In many cases, the supplementation of low-quality roughage by urea has led to increases in intake and rates of digestion (Campling, Freer, and Balch 1962).

Although the assessment of nutritive value by crude protein or digestible crude protein content is useful, the supply of dietary energy is important. Carbohydrate provides the main source of energy for the ruminant animal. Kingsbury (1965) suggested that pasture quality may be assessed in terms of soluble carbohydrate content and volatile fatty acid (V.F.A.) production. A high soluble carbohydrate content is often found in feeds of high digestibility although there is no direct correlation between these components. The soluble carbohydrate content of pasture grasses has a marked effect on the relative proportions of V.F.A. produced in the rumen. Grimes, Watkin, and Gallagher (1967) at Armidale showed that live-weight gain in sheep is positively correlated with soluble carbohydrate content in the grass and the proportionate fraction of V.F.A. in the rumen. Johns (1965) reported a negative correlation betweeen the cellulose content of herbage and live-weight gains in sheep. He considered that additional live-weight gains on clover as compared to grass may be due, in the main, to the more favourable V.F.A. pattern for fattening resulting from the high ratio of soluble carbohydrate to cellulose. There is thus some evidence that soluble carbohydrate may be as important as nitrogen content in determining the digestibility of a herbage and that high nutritive value may depend upon the attainment of a balance between these two components.

Chemical composition, digestibility, and intake

The relationship between voluntary intake and chemical composition and digestibility has been reviewed by Van Soest (1965). Six grass species and the legume lucerne were used in intake and digestibility trials with sheep. Lucerne had the highest intake and the highest lignin content. Tall fescue had the lowest intake and a low average lignin value. In grasses, the lignin content tends to increase more or less regularly over the season. In legumes the lignin content is higher than in grasses and its seasonal variation shows a less regular pattern, although there is an over-all increase with maturity.

Reid and Jung (1965) found that lucerne, in comparison with grass of equal digestibility, had a higher content of soluble carbohydrate and lignin, a lower cell-wall fraction, and a lower fibre digestibility.

Van Soest found confirmation for the proposition (Balch and Campling 1962) that fibre mass inhibits intake in those forages that have a high cell-wall content. The total fibrous part of legumes represented by cell-wall constituents does not appear to be large enough to inhibit intake. The point at which fibre mass becomes limiting occurs when cell-wall content lies between 50 and 60 per cent of the forage dry matter.

Hi Kon Oh, Baumgart, and Scholl (1966) examined 24 legume and 32 grass hays of which *in vivo* dry-matter digestibilities were known, to compare, as predictors of dry-matter digestibility, the contents of crude protein, acid-detergent fibre, acid-detergent lignin, cell-wall constituents, the solubility of cellulose in cupriethylene diamine (C.E.D.), and the solubility of dry matter in sulphuric acid (D.M.S.). All correlations between digestibility *in vivo* and laboratory tests were significant, but acid-detergent lignin was the best correlated of the chemical components studied ($r = -0{\cdot}60$ to $-0{\cdot}95$ for different species). On the other hand, Milford and Minson (1966) found no correlation between acid-detergent lignin and digestibility of Buffel grass and Rhodes grass.

Nutritive value index

In 1960 Crampton, Donefer, and Lloyd proposed the use of the nutritive value index (N.V.I.), in which the intake of digestible energy of feeds is expressed as energy digestibility (E) times the voluntary intake (I) in grams, the latter being adjusted to metabolic weight in kilogrammes:

$$\text{N.V.I.} = E \times \frac{I}{W^{0{\cdot}73}}$$

It was found that, as an approximation, the N.V.I. could be calculated from the twelve-hour *in vitro* digestibility (x) of the cellulose of the food:

$$\text{N.V.I.} = 1{\cdot}314x - 7{\cdot}8.$$

Lloyd *et al.* (1961) found the above equation to hold for a number of different herbage species. More recently, Johnson, Dehority, and Parsons (1965) found that the energy digestibility could be accurately predicted from C.E.D. × D.M.S., while relative intake was best predicted by D.M.S. *In vitro* cellulose digestibility values were more highly correlated with *in vivo* measurements for grasses while C.E.D., D.M.S., or both were more highly correlated with lucerne values.

It is obvious that relationships that may be used to predict the digestibility or intake of grasses cannot be expected without modification to serve the same purpose with legumes. Separate equations are required, and when these equations are applied to a mixed grass/legume sward the grass/legume ratio must be known.

Other methods of evaluation of pasture quality

Gallagher and Watkin (1967) endeavoured to use wool production and carcass quality in the evaluation of pasture quality by grazing sheep. No differences in wool growth were observed, nor was there any significant difference in carcass quality between treatments.

In our efforts to assess pasture quality on the basis of chemical composition we must not overlook the acceptability of the pasture to the grazing animal, since this influences intake; nor must we overlook the fact that the nutrient requirements for animal production are not uniform but depend on the type of production whether fattening, butter fat, milk, or wool production. Among chemical components of pasture, it would seem that more attention could with advantage be given to soluble carbohydrates and lignin.

APPENDIX A
The Ecological Basis of Grassland Husbandry

G. JACKSON
Lecturer, Department of Botany, University of Ibadan

GRASSLANDS are found in all parts of the world, providing pasturage and fodder for domestic and wild livestock. In general they are maintained to a greater degree by the users, whose aim is to keep the grassland in as productive a state as possible. Natural treeless grasslands are restricted to the arid regions of the world and are of low ground cover and carrying capacity. Elsewhere, grasslands are either wooded or are treeless because of some factor that eliminates tree growth. This implies that grass-dominated communities are not usually natural in terms of formation vegetation types.

A basic concept of plant ecology is that of the climax community. This is the vegetation that will predominate in an area in response to the prevailing conditions of climate and soils. It is the community where over-all equilibrium is demonstrated by numerical stability of numbers of individuals with no invasion by other species or elimination of the species present. The dominant plants in a climax community will be those forms that are able to make maximum use of the resources of the environment.

From a study of world climates and their associated vegetation types, it is clear that the greater the availability of moisture the larger the life forms that physically dominate the formation types. This is seen clearly in Africa, where the vegetation varies from desert to tropical rain forest with a large number of types filling in the range. The vegetation changes from grasses and herbs of sparse ground cover in the more arid regions to shrub-dominated steppe, through various degrees of wooded savanna to woodland, and finally in the equable tropics to rain forest. These changes correspond to an increasingly favourable water régime reflected in the increase in density and size of the woody perennials.

It is because grasses make less demand upon the habitat than trees that true climatic climax grasslands are found in areas of low rainfall efficiency. With more favourable growth conditions, grasses become sub-dominant and finally, in the rain-forest areas, are either sparse and specialized or even absent. It follows, therefore, that outside the arid zones treeless grasslands are either stages in succession to climax communities or are artificially maintained by factors that eliminate tree growth.

The factors that maintain grassland or divert the natural course of succession are usually biotic and more often than not under the control of man. Less often, edaphic factors are involved. Edaphic control indicates soil features that eliminate trees by poor drainage, high water table, or shallow soil conditions. African edaphic grasslands are predominantly those of the flood-plains and lake basins. Biotic factors are more important. Felling and farming are the more obvious causes of tree destruction, but fire is probably the most important single factor in widespread tree destruction. It might be said that climate, in terms of available moisture, controls the potential density and physiognomy of the tree cover whilst fire, together with cultivation and grazing practices, determine the actual density of the woody species.

In ecological terms, undisturbed woodlands are climatic climax types whilst communities in which the number of trees is less than that of climatic climax types are of a secondary nature and are derived or biotic climaxes, or edaphic climaxes controlled by soil conditions.

If a grassland and grazing programme is to be successful and grasses are to be improved or maintained in a productive state, the position of the grasses of an area in relation to the climatic climax community must be fully understood. Grassland agriculture is, of necessity, based upon the ecology of the grasses, especially where the grasslands represent sub-climax communities.

Grasses are, as a family, extremely successful. They play a part in most vegetation communities varying, in form, from ruderals and ephemerals to tall woody perennials and, in habitat, from near-desert to flood-plain and marshland. They have fibrous root systems which play a part in the building up of soil structure; the location of the growing point near the base of the plant enables regrowth after cropping; the habit of tillering gives rapid cover in early growth stages. These features of growth are closely related to habitat conditions of available moisture, nutrients, soil condition, light, and temperature.

Although widespread, and with major features in common, the principal tribes within the family tend to show well-defined ranges and areas of distribution. Many genera, however, may be described as cosmopolitan. The temperate tribes of Aveneae, Agrosteae, and Festuceae have a restricted distribution in the tropics and are mainly found on the high mountains. When members of these tribes are introduced into hotter areas they do not thrive. Andropogoneae and Paniceae, on the other hand, are predominantly tropical.

In grass-introduction programmes, the distribution of the species is taken into account, as it is not likely that grasses from one climatic zone will thrive outside their natural area unless the new habitat is climatically similar or cultivation practices compensate for the changed conditions. The transfer of grasses from their zone of natural distribution to similar climatic zones has been exploited. Many African grasses of the humid and arid tropical zones have been successfully cultivated in tropical America and Australia and vice versa. All such introductions have not always been deliberate. Some plants have been transferred accidentally from area to area and have become established under favourable conditions as troublesome weeds. *Panicum maximum* may, however, represent a species that has been accidentally carried from Africa to America, where it is now the main fodder for the Central American cattle industry.

When one considers grasses in relation to community dominants, it is seen that the grasses have a different and usually much wider range than the physiognomic forms used to designate the vegetation type. Thus *Monocymbium ceresiiforme*, for example, is found from the humid lowlands of West Africa to the high mountains of East and Central Africa. Within any vegetation type the grass may show a distinctly local distribution. *Monocymbium* in the humid tropics is a grass of shallow soils overlying rocks; in savanna areas it is a seasonal swamp dominant and, in the high mountains, it is a dominant of poor grassland and of shallow soils.

For efficient exploitation of grassland it is essential to understand the distribution of species and their autecology, taking into account the effect on the grass of habitat factors; the response to soil type, depth, drainage, moisture content, and nutrient status; the response to grazing, burning, and to other individual plants in the community; the effect of the varying climatic regimes; the range of the species, tolerance, amplitude, and adaptability. If such responses are understood, grassland maintenance and introduction can be based upon sound principles. It has been found that a grass such as *Chloris gayana* from somewhat saline lowland areas will thrive as a drought-resistant pasture grass in the central plateau areas of Africa. Similarly, other grasses of salt pans such as *Panicum coloratum* of the Makarikari pans and *Cynodon plectostachyus* are valuable pasture grasses under cultivation.

Study of the responses to biotic factors is possibly the most important aspect of grassland ecology, since it is only under the most primitive systems of cattle keeping that the grasses of natural climax communities are utilized. With the development of cattle keeping for profit it becomes imperative for grasslands to be made more productive. Grassland management and cultivation become part of the farming system and grasses are regarded as valuable crops.

Fire and its effects are most striking and have been the subject of much speculation and argument concerning their influence upon soils and soil erosion. What is certain is that fire through long, dry grass discourages tree growth. Certain fire-hardy species of tree survive burning when mature but succumb as saplings. Many fire-resistant trees have interesting young growth stages and have large, underground rootstocks. Grasses of the savanna appear to be fire-tolerant, and indeed some benefit from burning since fire destroys the long woody stems that otherwise remain from season to season. Grasses flush green in response to fire and draw upon root reserves before the onset of rain. The use of fire in pasture management must be based upon ecological principles if it is not to become an agent of destruction.

The effect of animals upon grasses is another biotic factor in grassland ecology. The grazing habits of certain animals encourage some grasses and eliminate others; animals spread weeds and tree seeds that can lead to pasture deterioration. Undergrazing and overgrazing have definite effects on plant successions and on the local distribution of species. The periodicity of grazing can affect composition by eliminating grasses grazed at susceptible stages in their seasonal growth.

Man plays a part in both the above biotic factors but he also consciously manages grasslands for productivity. Fire and the green grass flush are used by primitive pastoralists to give green fodder when fodder is otherwise scarce. Sophisticated grassland farmers use ecological principles to plan pasture management based on felling, burning, planting, fencing, rotation, and conservation.

Felling and judicious replanting of shade trees are practised since trees compete successfully with grasses and reduce yields. Burning is employed on a sound system to aid in bush eradication and the rejuvenation of the sward by removal of old trash. Fencing and rotation give opportunity for planned periodic grazing and resting. Conservation preserves food for adverse seasons. Planting is practised when indigenous pastures give low or uneconomic returns and where planted pastures are more productive than natural pastures. Planting assumes a knowledge of the behaviour under cultivation of the grasses used and involves the introduction of promising grasses from other areas.

Grasses show a wide range of tolerance and this is true of their relationship to soil types and conditions. The responses to conditions are expressed in terms of root development and vigour and correspond to the level of nitrogen, other nutrients, and water relationships. Poor aeration leads to a depression of root production in terms of both depth and lateral spread.

The marked response of grasses to climatic factors accounts for the restriction of species to zones. The temperature requirements for optimum growth vary from species to species and, there is evidence to show, from variety to variety derived from different climatic zones. Species vary in their water requirements and degree of drought-resistance. Growth, dormancy, and germination are correlated with climatic factors. Light intensity and day length influence development and flowering. The response of a grass to climatic factors is, therefore, limiting in programmes of introduction and development.

If grassland agriculture is to be understood and exploited to the full it is essential at the outset to understand the potentialities of grasses, their place in the natural communities, their habit, tolerance, range, and physiology—in short their ecological and biological relationships.

APPENDIX B
Grasslands of Tropical Africa

G. JACKSON

TRUE treeless grasslands are either climatic, edaphic, or biotic climax communities where habitat factors discourage the growth of trees.

Climatic climax grasslands occur in areas where rainfall efficiency is low and are associated with climates with low or ill-distributed rainfall, prolonged drought, low relative humidity, or with permanent or severe seasonal frost.

In Africa, climatic climax grasslands are confined to desert areas with prolonged drought, where rainfall is 12–25 cm annually, and erratic. Here there is a true treeless grassland with a sparse, open grass sward dominated by *Aristida* spp. and other Stipeae and members of the Chlorideae. The grassland is essentially short.

With a slight increase in wetness, desert scrub, with scattered shrubs of the genera *Commiphora* and *Acacia*, becomes the climax type.

Edaphic grasslands caused by soil conditions or high water table, and biotic grasslands in which trees have been eliminated by fire and grazing practices, are described in this appendix on p. 130.

The rest of African grasslands are in reality wooded grasslands in which the type of tree cover can be related to an increasingly favourable rainfall régime.

Grasslands in relation to climatic zones

In order of increasing rainfall we have the following sequence of formations described in terms of grass and woodland: desert; desert scrub; desert-grass, *Acacia* savanna; tall-grass, *Acacia* savanna; tall-grass, low-tree savanna; humid forests.

Desert grasslands in the 12–25 cm annual rainfall zone are found across the Sahara and in south-west Africa. Grass cover is often so low that the soil is unstable and the surface may blow away with the wind.

Desert scrublands, often called 'steppe', are also characterized, like the desert, by a large number of *Aristida* spp., although the flora is rich in many other species including *Enneapogon* spp., *Cenchrus ciliaris*,

Eragrostis spp., and *Panicum* spp. In the Somalia area *Chrysopogon aucheri* var. *quinquiplumis* assumes prominence.

The desert grasslands and scrubs are utilized by nomadic pastoralists whose animals, despite the apparent poverty of the vegetation, appear to thrive. This is probably a function of the drought, which does not cause leaching of the food materials in the dried-off herbage, which, like natural hay, is rapidly grown and rapidly dried off.

In the zone of 25–50 cm annual rainfall distributed through 4–7 wet months, there is a sparse sward of grasses with scattered thorn trees. The appearance has led to such descriptive terms as orchard-bush or thorn-bush steppe. The growing season is short, and the sparse grasses soon dry off and become brittle and yellow, following the end of the wet period. They are low-growing, and the area is of low carrying capacity in terms of livestock, but is excellent pasturage. Overgrazing of such savannas leads to the elimination of the grasses and increase in number of trees in a thicket form. This results in a general deterioration of the grazing potential, particularly in the neighbourhood of settlements.

The tree dominants are mostly *Acacia* spp., including *A. senegal* and *A. hebecladoides*, and members of the Capparidaceae and Combretaceae, *Commiphora* spp., and succulent *Euphorbia* spp. Many of the shrubs and trees are thorny or spiny, which has been interpreted as a protective mechanism against browsing. The grasses are typified by many *Aristida* spp. together with *Chloris* spp., *Chrysopogon* spp., *Enteropogon* spp., *Schmidtia bulbosa*, *Cenchrus biflorus*, and *C. ciliaris* and other low tufted types. These grasses are perennials, but there are often locally common annuals such as *Microchloa kunthii*, *Sporobolus festivus*, and *Tragus berteronionus*, which thrive during the rainy periods.

With increasing rainfall there is a corresponding increase in luxuriance of the vegetation and, in the zone of 50–100 cm annual rainfall, with 5 wet months, are found the tall-grass, acacia woodlands. These are amongst the most important grass–woodland associations in the continent, forming as they do a wide belt around the more humid woodland and forest areas.

Of the typical *Acacias* to be found in the zone are *A. spirocarpa*, *A. nigrescens*, *A. seyal*, *A. nilotica*, and *A. senegal*; together with *Sclerocarya birrea*, *Tamarindus indicus*, *Burkea africana*, *Dichrostachys glomerata*, *Combretum* spp., and *Terminalia brownii*. The grasses are tall and tufted and include the familiar *Hyparrhenia* spp. and *Andropogon* spp.

With further increase in rainfall, in the 88–150 cm zone, with 7 wet months, there is a change to the tall-grass, low-tree woodland, or savanna woodland. 'Low tree' is an unfortunate term since the 'low trees' are often quite tall and include *Daniellia oliveri* and *Butyrospermum parkii*. The grasses are tall and often very coarse, and include *Hyparrhenia chrysargyrea*, *H. rufa*, *H. dissoluta*, *H. subplumosa*, *Schizachyrium sanguineum*, *Ctenium nubicum*, *Andropogon schirensis*, and *A. gayanus*. Because of the reedy nature of these grasses when mature, the problem of grazing these areas is one of encouraging the finer grasses such as *Digitaria* spp., *Setaria* spp., *Panicum dregeanum*, and *Brachiaria* spp., which generally appear in the early wet season and then are overtopped by coarser species.

It is the area of fierce bush fires, and the trees show the characteristics of perennials able to withstand annual burning, such as corky fissured bark, deciduous habit, coppice and sucker regrowth, and superficial and extensive root systems.

The tall-grass, low-tree zone gives way to the forest zones as humidity and rainfall rise.

In summary, there is a gradual increase in size both of trees and grasses with increasing rainfall, until grass–woodlands give way to forest proper, where grasses are specialized or absent. Low-rainfall areas are characterized by low, quick-growing, tufted types of low ground cover; high-rainfall areas have tall, coarse grasses, still tufted but with greater vegetative cover. Emphasis should be laid here on the fact that actual ground cover is low even in the higher-rainfall zones. This may be a consequence of the dry season and the competition for moisture for rapid growth at the onset of the rains.

Life form of grasses is a reflection of climatic conditions. It is equally well seen to be a reflection of edaphic conditions in the study of grasses in seral communities.

Grasses in seral communities

In nature, any bare surface of land is sooner or later colonized by plants; at first by simple life forms and later, as reaction between the plants and the habitat takes place, by higher forms as conditions for plant life improve. On rocks, this reaction is generally an accumulation of soil and organic debris, which is a consequence of chemical and physical weathering of the rock and a degree of stabilization by plants. It is aided and accelerated by cracks and fissures in the rocks, which fill with debris and form suitable substrata or media for plant growth. Early in the succession of such areas one finds grasses. Their demands on soil depth are low, and they will be found established either on the shallow margins of plant mats or in narrow and shallow cracks. These grasses, occupying shallow soils, are usually annuals of small stature, which survive from season to season by setting copious quantities of seed.

Panicum griffonii, a larger annual, is found upon slightly deeper soils. During the rains it is able to form quite a good top cover, not of great height but fairly dense vegetatively. It has brittle and fine stems and sets a great deal of light seed. Other annuals of similar situations include *Euclasta condylotricha* and *Diectomis fastigiata*, both forming a tangled vegetative cover with long brittle stems. *Ctenium nubicum*, in its annual form, is a tufted fine-stemmed grass found on shallow soils or in rock crevices.

With gradual increase in soil depth, annuals are replaced by perennials of medium stature, such as *Andropogon linearis*, *Loudetia arundinacea*, *Monocymbium ceresiiforme*, and perennial *Ctenium nubicum*.

In the final grass stages the taller coarse perennials become dominant in the communities that precede colonization by woody plants, and the wetter the region the taller and more robust the grasses of this stage are likely to be.

In East Africa, *Chloridion cameronii* is frequently found where one would expect to find *Euclasta* and *Diectomis* in Nigeria. This is followed by *Loudetia simplex*, *Monocymbium ceresiiforme*, and *Themeda triandra*. An interesting fact is that *M. ceresiiforme* occupies only shallow soils in Malawi, if in a montane, high-rainfall zone. Elsewhere it is confined to swampy grasslands.

On very shallow soils and indurated laterite outcrops *Microchloa kunthii* is a common pioneer and many annual *Sporobolus* spp. are found in similar situations.

The foregoing are the grasses of the particular primary succession known as a *xerosere* (on rock surfaces a *lithosere*), where increasing size and increased life span indicate an improvement in habitat conditions.

At the other end of the scale is the *hydrosere* or succession from open water. Here the form of the plants reflects the excess water, and the types involved are generally perennials. The tropical hydrophytic grasses are numerous and spread through many genera and tribes. Some of the more important species involved are given below.

At the outermost fringes of vegetation colonizing open water are the submerged and free-floating plants, which are non-grasses, generally *Ceratophyllum* spp., *Vallisneria*, *Utricularia*, and *Pistia*. These are followed by the floating-leaved plant forms such as *Nymphaea*, *Eichornia cordifolia*, and *Trapa natans* as the water gets shallower. It is in the next, shallowest, zone that the grasses and sedges become dominant. Their life form is one of floating stems with leaves and flowering culms above the water, or of rhizomes rooted in the mud with upright stems growing up and above the water level. The floating grasses are naturally suited to the deeper water areas. Typical of this group is *Vossia cuspidata*, common along the lagoons. The zone in which it flourishes stretches from the shallow waters of the shore, where it may be firmly rooted in the mud, out into the water where the grass is free-floating. Associated with it in many parts of Africa is the grass *Echinochloa pyramidalis* of very similar habit and life form. It has floating stems with broad, long, lanceolate leaf blades carried above the water. Both *Vossia* and *Echinochloa* can, under certain circumstances, form floating mats capable of carrying a person. They are also found together with the papyrus sedge (*Cyperus papyrus*) and form the typical sudd of Africa's major rivers and large swamps. There are smaller grasses of similar habit associated with swamp conditions, such as *Acroceras amplectens*, *A. zizanioides*, and *Leptochloa coerulescens*, which often form mats on small bodies of water.

In shallower water (but sometimes assuming the habit of *Vossia*), are the grasses that are rooted in the mud and carry their stems above water. *Phragmites mauritianus* is a good example. This is a tall reedy grass, which forms a dense reed-swamp type of vegetation. It is usually found where there has been an accumulation of mud so that, at times of water retreat, the soil surface is above water. If the water remains deep for any length of time, *Phragmites* does develop long stems that float on the water surface. In damp, semi-swamp areas it can become a pernicious weed, difficult to eradicate because of its rhizomatous nature.

Similar in form is *Pennisetum purpureum*. This grass has a wide range of habitat and is often associated with land reverting to forest after cultivation, but it is more common in wetter places. Its natural distribution in Africa is a guide to its high demands upon water supplies. In Malawi it is confined as an upland grass to those districts with dry-season mists. It should be noted that under cultivation it is a valuable pasture type but, if neglected or subjected to competition, it becomes reedy and less leafy and finally dies out.

Saccharum spontaneum, the wild sugar-cane, is of similar form and habitat.

There is a wide range of grasses associated with swampy, rather than permanently flooded, land. These grasses are generally of a creeping habit and form a well-dispersed vegetative cover. They include *Paspalum conjugatum*, *Leersia hexandra*, and *Chloris gayana*. As the water retreats during the dry weather, many areas of such grassland give very reasonable pasture to livestock.

In the savanna zones with a marked dry season there are the seasonal swamps of valley bottoms. These carry a well-dispersed grassland with tufted grasses, such as *Hyparrhenia* spp., together with many finer grasses in the bottom of the sward. This type of grassland has a distinct structure.

Although one tends to refer to seasonal swamps as edaphic grasslands where trees are eliminated due to a high water table, it is much more likely to be fires in the dry season that are the cause of treelessness. From seasonal swamp and seasonally flooded grasslands the sere proceeds into swamp woodland or woodland proper, according to the circumstances.

Edaphic grasslands are not permanent in the sense that, if left undisturbed, sooner or later woodland will be the result of reaction and succession.

Grasses in secondary successions

Grasses play a great part in the subsere or secondary succession that follows the abandonment of cultivation. Neglected cultivations on newly abandoned farms carry a complex of weeds in which are included such annual grasses as *Eragrostis ciliaris*, *E. tenella*, *E. aspera*, *Panicum kerstingii*, *Chloris pilosa*, *Digitaria horizontalis*, and *Setaria barbata*.

Following the low-growing short-cycle annuals are the larger annuals, the particular ones found in an area reflecting the probable fertility status of the lands invaded. Soils of low fertility are colonized first by *Rhynchelytrum repens* (red top grass), a loose, creeping and straggling annual with silky-hairy spikelets that are wind-dispersed. It is followed in the succession by *Pennisetum polystachyon*, a profusely seeding annual of tufted habit. In the later annual stages the tall annual *Rottboellia exaltata* may be found. It is interesting to note that on more fertile soils the *Rhynchelytrum* stage may not take place, and *Pennisetum* is the first prominent invader. The vigour of *Rottboellia* is a useful indication of soil fertility status. The annuals are succeeded by perennial grasses, e.g. *Andropogon* and *Hyparrhenia* spp., according to the climatic zone.

This is by no means the whole story. There are other grasses of the seral stages, but these examples serve to show the gradual change in form and habit consequent on the progress of a sere.

In the case of subseres, the initiation of development is not on a new bare area but where there has been destruction of an original vegetative cover. The soil is thus well or fully developed and the time taken for a subsere to reach completion will be very short compared with a primary succession from bare rock or water. The actual stages in the subsere and the time of completion will depend, however, upon the treatment the bared soil has received following the destruction of the original cover.

Forestry and felling create least disturbance since they simulate the natural fall of tree vegetation, and, although the area of destruction is much larger, the soil is little disturbed. The subsere will thus be initiated at the evergreen forest tangle stage if in the forest zone.

Farming activities stir and disturb the soil, break down structure, and deplete the nutrient status. The degree of damage to soil structure and the loss of fertility will control the stages in succession and the length of time it takes to completion. Farming activity that results in erosion and actual removal of soil will lengthen the subsere, and even drastic soil movement may result in bare surfaces and a prisere will then be initiated by nudation; this is so particularly where indurated laterite is found near the soil surface.

At this point it is of interest to note the probable role of grasses in the subseral stages and the short-term changes that are observed following farming activities. Cultivation results in the loss of soil structure as the peds or aggregates are disturbed and broken down; water relationships deteriorate unless a careful organic-manuring régime is adopted; fertility is lost through removal of nutrients; some topsoil is lost through erosion and the topsoil tends to become dusty and, in consequence, impedes infiltration of rain water.

It has been found that the depth of the root systems of grasses can be related to the pore space and the structure of the soil (Martin 1944). Since pore space is reduced and structure broken down during cultivation, the pioneers in the subsere are likely to be the more shallow-rooted annual grasses. As the root systems of these grasses grip and bind soil particles, and help to restore the soil structure, so the deeper-rooted annuals and perennials succeed. It is thought that it is in this manner that the grasses are valuable in restoring fertility by improving soil structure and increasing porosity. Grasses do not fix nitrogen but, depending upon the depth and extent of their root systems, tap soil resources and contribute in the long run to increased nutrient status of the topsoil. They contribute considerable amounts of organic matter. Grasses do not restore soil lost by erosion but their physical nature, particularly when of dense cover, helps to reduce wash, holds up silt brought from elsewhere and reduces the impact of rain on the soil surface. Grasses are thus of supreme importance to man, to agriculture, and to conservation practices.

That soil structure is of prime importance in subseral succession, and is correlated with the development of grass cover, is evidenced by the difficulty that is experienced when attempting to establish grass swards in areas of worn-out soils, or where severe trampling and compaction have taken place and tall grasses are completely discouraged. There are, however, certain wild species of grass such as *Eleusine indica* and *E. africana* that do thrive in trampled areas and become dominant along cattle tracks and paths.

The position of grassland in succession

All the above examples point to the fact that treeless grasslands, with the exception of climatic grassland, are seral stages in plant succession, giving way under a natural sequence of events to woodland or forest communities. We shall now therefore look at the factors that may influence the prolongation of the grassland stages in an area where the natural climax is woodland.

Edaphic factors are in the main responsible for natural grasslands in woodland areas, particularly those factors concerned with the water table. In areas such as wide valley bottoms or lake basins, where there is prolonged flooding or waterlogging (and particularly where there is a cool season), there is a tendency to find grassland communities. This is the case in the savanna regions where the wide seasonal swamps are a characteristic feature of the landscape and where such grassland bottoms are valuable assets to grazing people; but, before one concludes that such seasonally flooded areas are true grasslands, it is as well to mention that *Acacia campylacantha* has been found to invade such areas where cutting rather than burning of grass is practised.

Climatic climax grasslands are found at high altitude, on mountains above a line beyond which tree vegetation is no longer found, temperature being the controlling factor.

At slightly lower altitudes, of about 1800–2100m, there are extensive grasslands where forest has been destroyed in the past and where no other woody species are present to replace the destroyed species. The controlling factor here is fire, and thus montane grassland for the most part can be regarded as biotic.

Biotic factors, mainly the effect of grazing and fire, are responsible for the maintenance of grasslands in forest and woodland areas. Fairly fierce grass fires, usually of annual occurrence in much of Africa, are used by pastoralists to encourage a green flush of grass during the drought period. Grass in the savanna areas reacts to the burning by putting out new shoots after the fire, drawing on reserves of moisture and food in the rhizomes and root systems and naturally depleting such reserves prior to the onset of the rains. If the grass is then defoliated severely by grazing there is a tendency for the tufts to be weakened and to deteriorate and they may even be killed and replaced by annual or unpalatable grasses.

Burning also destroys or burns off young coppice growth and tree seedlings, particularly if fierce. Relaxation of burning can thus lead to a thickening up of woody vegetation in a grass-woodland area. This, of course, is what happens when grazing reduces the bulk of grass available for firing, so that fires are not fierce enough to check woody species. In *Acacia* areas this leads to thicket formation and deterioration of grazing potential.

Fire has the deleterious effect of exposing the soil surface during periods of high insolation, and this is thought to have an adverse effect on soil.

Elimination of fire altogether has been found, in the purple-veld areas of South Africa, to cause a deterioration of the grassland and eventually to lead to a poor sward of annuals. Thus fire appears to be a necessary factor in the maintenance of grassland but, at the same time, can be a destructive and dangerous agent. It will suffice here to say that used in a sensible way fire is useful, especially if employed only once in several years. Under such circumstances fire will clear up the old accumulated dead stems of the grasses, which are unpalatable and very resistant to breakdown, will prevent, by destruction, scrub formation or bush encroachment, set back the coarse tufted grass species, and give seedlings of other grasses a chance to grow early in the season if they are vigorous enough to become established.

Grazing

This is the second main biotic factor. It is either practised deliberately by man or naturally by herds of game. Its effects are varied. Overgrazing occurs where there are too many cattle in an area and the carrying capacity is exceeded. Grasses are heavily defoliated and, since a defoliated grass is unable to produce a healthy root system, the result is deterioration and death and replacement in the sward by grasses that are either not so demanding or are less palatable to grazing animals. The order of replacement is as follows: tufted Andropogoneae are replaced by creeping perennials and free-seeders and these, in turn, by weeds or by unpalatable species. If the succession can be halted at an intermediate stage a very desirable grassland results. But such grassland is often the result of a very delicate balance.

Understocking without control, on the other hand, will lead to a patchy grassland with an uneven sward, since the areas first grazed in the season will tend to be overgrazed, because the young grass that results after defoliation is much more palatable than the ungrazed tufts that have grown and become stemmy. It is no uncommon sight to see overgrazing and undergrazing in an understocked, uncontrolled area. Undergrazing leads eventually to very poor grassland and to bush encroachment.

A good farmer or pastoralist will base the use of his grazing lands on sound ecological principles if he wishes to maintain them as good pasture. Paddocking (in fenced or beaconed grazing areas) with rotational grazing is practised to reduce the effects of selective grazing; correct stocking rates, together with conservation of wet-season surpluses, eliminate the risks of overgrazing and undergrazing; and rotational burning prevents bush encroachment. Rotational systems also allow for resting periods, which enable the grasses to replace root reserves and to set seed. Resting grassland after burning is a wise practice.

Seasonal development

In grassland there are distinct phases of seasonal development, at which times the casual observer might be tempted to identify and classify the grassland by its seasonal aspect.

Thus the *Digitaria*, *Setaria*, and *Panicum* spp. tend to flower first and be conspicuous early in the growing season. *Hyparrhenia* and *Andropogon* flower last in the season, and the middle phase is dependent upon area and vegetation type.

Life form and habitat factors

We now return to the types of grasses found, and their relationship to habitat. There are, broadly speaking, six main types, tufted, creeping, scrambling, woody, annual, and the specialized types found in evergreen forests.

Tufted grasses maintain close tufts, and the resulting grassland is of low ground cover, possibly because of the extensive fibrous root systems and intensive inter-tuft competition. They are typical of the savanna woodlands and mountain grasslands.

The size of the stem varies. In the tall-grass, low-tree savanna the grasses tend to be tall and the stems or culms become tough and reedy. In the montane zones the grasses are low-growing and much more leafy, with a close vegetative cover, but a walk in mountain areas soon reveals the tufted nature of the sward and that the actual ground cover is low.

Tufted grasses may branch intra- or extra-vaginally. In the closely-tufted types the axillary bud grows without breaking through the leaf-sheath. In this case, the grass is said to branch intra-vaginally. In loosely tufted grasses branching may be extra-vaginal, in which case the axillary bud breaks through the leaf-sheath near the base.

Creeping grasses generally have extra-vaginal branching and spread by stolons or by rhizomes. The ground cover tends to be denser than that of the tufted types and creeping grasses are often associated with better water supplies. *Chloris gayana* affords a good example of a creeping stoloniferous grass of marshy areas. The stolons loop along from the parent plant and root at the nodes, each node producing a new plant. In fairly wet areas the internodes are short and ground cover is dense, but in drier areas the internodes tend to be longer and, under such conditions, a tufted sward of low ground cover may result.

Scrambling grasses tend to grow at the edge of and amongst bushes and woody regrowth. Their stems are often very thin and brittle and may be scabrid. Typical of this type are many of the Melineae, *Andropterum stolzii*, *Hyparrhenia cymbaria*, and *Ehrharta abyssinica*.

There are many such types at the edges of montane forest and along the paths and regrowths in the wetter forest regions. There is an *Acroceras* sp. in Cameroon with this habit, and several *Panicum* spp.

Woody grasses. There are several grasses with woody persistent stems, of which bamboos are typical examples. *Olyra latifolia* of the forest approaches this type, whilst bamboos, such as the indigenous *Oxytenanthera abyssinica* and *Arundinaria alpina*, are capable of forming pure-stand communities.

Annuals are very numerous and successful members of the grass family. They are the pioneers in primary and secondary successions.

Evergreen forest types are often of specialized form. The leaf blades tend to be broad and thin and often have acuminate tips; many have falsely petiolate leaf blades, and the blade is held upside down by virtue of a twist in the false petiole. Their life form fits into the categories listed above but, if tufted, the tufts tend to be loose and open; if creeping, the stolons or rhizomes are loosely rooted in the surface of the forest floor and there is little tendency to form a tight mat.

Distribution of grasses

Grasses can be said to be some of the most successful of terrestrial plants. They are found in all types of habitat and under many conditions.

Their life form varies from woody perennials to ephemeral annuals varying with habitat. Many have xeromorphic features that enable them to survive under drought conditions and they are the dominants of the dry areas, many of them showing special features.

The most primitive of the grasses are thought to be the bamboos, which are elevated by some workers to a sub-family, the Bambusoideae. They have generally six stamens and a woody perennial habit, and are largely confined to the tropical areas where the climate is equable and humid; some are hydrophytes and even aquatics. Africa is poor in bamboos but has species in more specialized habitats, that is *Oxytenanthera abyssinica* of the dry escarpments and *Arundinaria alpina* of the mountains.

From this group, workers have seen the possible derivation of the Panicoideae.

Tropical African grasses are predominantly of the sub-family Panicoideae and, in fact, the two tribes Andropogoneae and Paniceae account for the vast majority. There is a tendency for the Andropogoneae to be the dominant grasses of the climax and sub-climax wooded grasslands, whilst the Paniceae are represented by the grasses of more specialized habits such as the ruderals and hydrophytic types, although many exceptions are found. The Paniceae and the Andropogoneae have a few representatives outside the sub-tropics, such as the North American species of *Panicum.*

Genera of the Stipeae, such as *Aristida* and *Stipa*, are characteristic of climatic grassland types. Elsewhere in the tropics they are found as ruderals and annuals: for example, overgrazing and overburning of Andropogoneae grasslands lead to predominance of *Aristida* spp. in the drier areas.

Many Chlorideae are found in dry areas; even *Chloris gayana*, a swamp grass, may be subject in the dry season to prolonged drought.

The Eragrosteae and the closely related Pappophoreae and Sporoboleae are typical of the drier areas and have many ruderal representatives.

The Arundinelleae are grasses of dry areas in woodlands or are found on sandy, seasonal-swamp soils.

The Oryzeae and the Arundineae are hydrophilic grasses.

The predominantly temperate tribes of grasses are the Festuceae, the Hordeae, the Aveneae, and the Agrosteae. They are poorly represented in the hotter regions but are plentiful in the montane grasslands of Africa, many temperate genera and some species being well represented in both Europe and Africa.

Bibliography

General references

AHLGREN, G. H. (1949). *Forage crops.* McGraw-Hill, New York.

ARCHER, S. G. and BUNCH, C. E. (1953). *The American grass book: a manual of pasture and range practices.* Oklahoma.

DAVIES, W. (1958). Grassland developments in the United Kingdom in the last 25 years. *Emp. J. exp. Agric.* **26**, 195–207.

—— (1959). *The grass crop.* 2nd edn, Spon, London.

—— (1960). Pastoral systems in relation to world food supplies (Address to British Association, Section M, Sept. 1960). *Rep. Br. Ass. Advmt Sci.* **67.**

HARLAN, J. R. (1956). *Theory and dynamics of grassland agriculture.* Van Nostrand, New York.

HUGHES, H. D., HEATH, M. D., and METCALFE, D. S. (1951). *Forages: the science of grassland agriculture.* Iowa State College Press.

LEVY, E. B. (1955). *Grasslands of New Zealand.* Government Printer, Wellington.

MOORE, H. I. (1949). *The science and practice of grassland farming.* Nelson, London.

—— (1950). *Grassland husbandry.* 4th edn, Allen and Unwin, London.

U.S.D.A. (1948). *Grass: the yearbook of agriculture.* Government Printing Office, Washington.

VOISIN, A. (1959). *Grass productivity.* Crosby Lockwood, London.

WATSON, S. J. (1951). *Grassland and grassland products.* Arnold, London.

WHEELER, W. A. (1950). *Forage and pasture crops.* Macmillan, New York.

WHYTE, R. O., MOIR, T. R. G., and COOPER, J. P. (1959). *Grasses in agriculture.* (Studies in Agriculture no. 42.) F.A.O., Rome.

Chapter 1

Text reference

HAMILTON, R. A. (1955). *Agriculture, Lond.* **62,** 374, 407.

Suggested further reading

CURRIE, J. R. (1948). The economics of ley farming. *J. Br. Grassld Soc.* **3,** 27–34.

Report of the Committee on Grassland Utilization (1958). Cmd. **547,** H.M.S.O., London.

Chapter 2

Text references

ARMSTRONG, S. F. (1937). *British grasses.* C.U.P., London.

COBLEY, L. S. (1956). *Introduction to the botany of tropical crops.* Longman, London.

GRIFFITH, G. (1949). *E. Afr. agric. J.* **14,** 187–8.

HARTLEY, W. and WILLIAMS, R. J. (1956). *Proc. 7th Int. Grassld Congr.* **191.**

HUBBARD, C. E. (1959). *Grasses.* Penguin, Harmondsworth.

MARTIN, W. (1944). *E. Afr. agric. J.* **9,** 189–95.

Chapter 3

Text references

BURTON, G. W. (1951). *Adv. Agron.* **3,** 201–5.

IVINS, J. D. (1952). *J. Brit. Grassld Soc.* **7,** 43–54.

MINSON, D. J., RAYMOND, W. F., and HARRIS, C. E. (1960). *Proc. Int. Grassld Congr.* 470.

ROGLER, G. A. (1944). *J. Am. Soc. Agron.* **36,** 487–96.

Suggested further reading

BOGDAN, A. V. (1949). Observations on palatability of some leguminous plants of Kenya. *E. Afr. agric. J.* **25.**

MELVILLE, J. (1953). Pasture quality and animal production. *Proc. N.Z. Soc. Anim. Prod.* **13,** 65–73.

SYNGE, R. L. M. (1952). The utilization of herbage protein by animals, *Br. J. Nutr.* **6,** 100–4.

WEINMANN, H. (1948) Underground development and reserves of grasses. *J. Br. Grassld Soc.* **3,** 115.

Chapter 4

Text references

HEGARTY, M. P. and POUND, A. W. (1968). *Nature, Lond.* **217,** 354.

HUTTON, E. M. (1962). *Aust. J. exp. Agric. Anim. Husb.* **2,** 117.

JOZWICK, F. X., NICHOLLS, A. O., and PERRY, R. A. (1970). *Proc. 11th Int. Grassld Congr.* (*Queensland*) 48.

WHYTE, R. O., MOIR, T. R. G., and COOPER, J. P. (1959). *Grasses in Agriculture* (Studies in Agriculture no. 42). F.A.O., Rome.

WHYTE, R. O., NILSSON-LEISSNER, G., and TRUMBLE, H. C. (1953). *Legumes in Agriculture* (Studies in Agriculture no. 21). F.A.O., Rome.

Suggested further reading

BOGDAN, A. V. (1954). Herbage plants at the Grassland Research Station, Kitale, Kenya, *E. Afr. agric. J.* **20,** 151–65.

—— (1956). Indigenous clovers of Kenya, *E. Afr. agric. J.* **22,** 40.

BRYAN, W. W. (1969). *Desmodium intortum* and *D. uncinatum*. A review. *Herb. Abstr.* **39,** 183.

CLAYTON, W. D. (1960). A key to Nigerian grasses. *Samaru Res. Bull.* **1.**

EDWARDS, D. C. (1940). Pasture and fodder grasses of Kenya. *E. Afr. agric. J.* **5,** 248.

—— and BOGDAN, A. V. (1951). *Important grassland plants of Kenya.* Pitman, Nairobi and London.

HALL, M. (1948). Five hundred varieties of herbage and fodder plants. *Bull. Commonw. Bur. Past. Fld Crops* **39.**

HORRELL, C. R. (1958). Herbage plants at Serere Experimental Station, Uganda, 1954–7; (ii) Legumes. *E. Afr. agric. J.* **24,** 133–8.

HUBBARD, C. E. (1926, 1927). *East African pasture plants.* Crown Agents, London.

—— and VAUGHAN, R. E. (1940). *The grasses of Mauritius and Rodriguez.* Crown Agents, London.

JACKSON, G. and WIEHE, P. O. (1958). *An annotated check list of Nyasaland grasses, indigenous and cultivated.* Government Printer, Zomba, Nyasaland.

MICHEL, G. (1960). The forage plants of the Belgian Congo. *Bull. agric. Congo belge* **51,** 567–602.

RAINS, A. B. (1956). A field key to the commoner genera of Nigerian grasses. *Niger. Fld* **22,** no. 4.

SENARATNA, S. D. J. E. (1956). *The grasses of Ceylon* (Peradeniya Manual no. 8). Ceylon Department of Agriculture.

STRANGE, R. (1954). Legumes for medium altitudes in Kenya. *E. Afr. agric. J.* **20,** 221–4.

Chapter 5

Text references

BEDDOWS, A. R. (1944). *Bull. Commonw. Bur. Past. Fld Crops* **31.**

BLAIR, J. J. (1960). *J. Br. Grassld Soc.* **15,** 206–51.

BULLER, R. E. (1960). *Proc. 8th Int. Grassld Congr.* 374.

CATFORD, J. R. (1952). *E. Afr. agric. J.* **17,** 183–7.

CHRISTIAN, C. S. and DONALD, C. M. (1949). *The Australian environment.*
DABADGHAO, P. M. (1960). *Proc. 8th Int. Grassld Congr.* 226.
DAVIES, W. (1959). *The grass crop.* 2nd edn, Spon, London.
FROMENT, D. (1959). *Proc. 3rd Inter-Afr. Soils Conf., Dalaba.*
HENKEL, J. S. (1931). *Proc. Trans. Rhod. scient. Ass.* **30.**
HITCHCOCK, A. S. (1927). *Contr. U.S. natn Herb.* **24,** 291–556.
KEAY, R. W. J. (1959). *An outline of Nigerian vegetation.* 3rd edn, Government Printer, Lagos.
LEOPOLD, A. S. (1950). *Ecology* **31,** 507–18.
LEVY, B. (1955). *Grasslands of New Zealand.* Government Printer, Wellington.
MCTAGGART, A. (1935). *Herb. Rev.* **3,** 136–51.
MOORE, R. MILTON, ed. (1970). *Australian grasslands.* Australian National Press, Canberra.
MOTTA, M. S. (1956). *Proc. 7th Int. Grassld Congr.* 539.
PERRY, R. A. (1960). *Land Res. Ser. C.S.I.R.O. Aust.* **5,** 1.
PITTIER, H. (1942). *The mesa of Guanipa.* Caracas.
RAINS, A. B. (1956). *Niger Fld* **22,** 4.
RATTRAY, J. M. (1960). *The grass cover of Africa* (Studies in Agriculture no. 49). F.A.O., Rome.
ROSEVEARE, G. M. (1948). The grasslands of Latin America. *Bull. Commonw. Bur. Past. Fld Crops* **36.**
SNOW, O. W. (1948). *Agriculture in the Sudan* (ed J. D. Tothill). O.U.P., London.
SPECK, N. H., BRADLEY, J., LAZARIDES, M., PATTERSON, R. A., SLATYER, R. O., STEWART, G. A., and TWIDALE, C. R. (1960). *Land Res. Ser. C.S.I.R.O. Aust.* **4.**
TATON, A. (1949). *Bull. agric. Congo belge* **40,** 1884–1900.
THOMAS, A. S. (1940). *Agriculture in Uganda* (ed. J. D. Tothill). O.U.P., London.
TRAPNELL, C. G. (1953). *The soils, vegetation and agriculture of N.E. Rhodesia.* Government Printer, Lusaka.
TRAPNELL, C. G., MARTIN, J. D., and ALLAN, W. (1950). *Vegetation and soil map of Northern Rhodesia.* Government Printer, Lusaka.
VAN RENSBURG, H. J. (1960). *Proc. 8th Int. Grassld Congr.*
WHITE, C. T. (1934). *Herb. Rev.* **2,** 5–7.
WHYTE, R. O. and RATTRAY, J. M. (1958). *Nature, Lond.* Jan. 11, 1958.

Suggested further reading

CROWDER, L. V. (1967). Grasslands of Columbia. *Herb. Abstr.* **37,** 237.

DEVRED, R. (1956). The savanna herbage of the Mouazi Region (Lower Congo). *Bull. Inf. I.W.E.A.C.* **65.**

GROSSMAN, J., ARONOVICH, S., and CAMPELLO, E. DO C. B. (1965). Grasslands of Brazil. *Proc. 9th Int. Grassld Congr.* (*Brazil*) **2,** 11.

JOTTRAND, M. and HENRIOUL, R. (1960). Dairy farming in Katanga (Congo). *Bull. Inst. agron. Stns Pech. Gembl.* **3** (hors sér), 1333–78.

MATHIEU, P. (1960). Livestock rearing in Urundi (Central Africa). *Bull. agric. Congo belge* **51,** 885–922.

MOTTA, M. S. (1952). *Emp. J. exp. Agric.* **22,** 283.

ROTAR, P. P. (1968). *Grasslands of Hawaii.* University of Hawaii Press.

THOMAS, A. S. (1942). Ecological factors and indicator plants in tropical Africa. *Chronica Bot.* **7.**

TROCHAIN, J. L. and KOESHLIN, J. (1958). The natural pasture in the south of Equatorial Africa. *Bull. Inst. Étud. centrafr.* **15–16,** 59–83.

VAN RENSBERG, H. J. (1967). Legumes and grasses in Zambia. *Herb. Abstr.* **37,** 315.

WHYTE, R. O., MOIR, T. R. G., and COOPER, J. P. (1959). *Grasses in Agriculture* (Studies in Agriculture no. 42). F.A.O., Rome.

Chapter 6

DAVIES, W. (1960). *8th Int. Grassld Congr.* presidential address.

Chapter 7

Text references

BASHAW, E. C. (1961). *Agron. Abstr.* **47.**

HUTTON, E. M. (1964). *Bull. Commonw. Bur. Past. Fld Crops* **47,** 79.

HUTTON, E. M. (1965). *J. Aust. Inst. agric. Sci.* **31,** 102.

—— (1968*a*). *Proc. Aust. Grassld Congr.* (*Perth*) **49.**

—— (1968). *Span* **11,** 73.

LUDLOW, M. M. and WILSON, G. L. (1967). *Aust. J. agric. Res.* **19,** 35.

WHYTE, R. O., MOIR, T. R. G., and COOPER, J. P. (1959). Grasses in Agriculture (Studies in Agriculture no. 42). F.A.O., Rome.

Suggested further reading

BOGDAN, A. V. (1966). Plant introduction, selection, breeding and multiplication. *Tropical pastures* (ed. W. Davies and G. L. Skidmore), p. 75. Faber and Faber, London.

DAVIES, W. (1959). *The grass crop.* 2nd edn, Spon, London.

WHEELER, W. A. and HILL, D. D. (1957). *Grassland seeds.* van Nostrand, Princeton, New Jersey.

Chapter 8

Text references

BUNTING, A. H. and LEA, J. D. (1956). *Emp. J. exp. Agric.* **24,** 95, 185.
DAVIES, J. G. (1960). *Proc. 8th Int. Grassld Congr.* Paper 16 A/5.
JOHNSTONE-WALLACE, D. B. (1937). *Proc. 4th Int. Grassld Congr.* 188–96.
MILES, J. F. (1949). *Rep. Commonw. scient. ind. Res. Org.* **6.**
MOORE, A. W. (1960). *Nature, Lond.* **185,** 638.
SEARS, P. D. (1953). *N.Z. Jl Sci. Technol.* **35,** suppl. 1.
VAN KEUSEN, R. W. and HEINEMANN, W. W. (1958). *Agron. J.* **50,** 85–8.
VASQUEZ, L. N. (1957). *Agric. trop. Bogotá* **13,** 369–71
—— (1958). *Herb. Abstr.* **28,** 39.
VICENTE-CHANDLER, J., CARO-COSTAS, R., and FIGARELLA, J. (1953). *Agron. J.* **45**, 397.
WALKER, T. W., ORCHISTON, H. D., and ADAMS, A. F. R. (1954). *J. Br. Grassld Soc.* **9,** 249–74.
WALSH, S. R. (1958). *Qd agric. J.* **84,** 527–36.
WARMKE, H. E. *et al.* (1952) *Trop. Agric., Trin.* **29,** 115–21.
WHITNEY, A. S. (1968). *Herb. Abstr.* **38,** 143.
—— and KANEHIRO, Y. (1967). *Agron. J.* **59,** 585.

Suggested further reading

BLASER, R. E. and BRADY, N. C. (1950). Nutrient competition in plant associations. *Agron. J.* **42,** 128–35.
BLASER, R. E., SKRDLA, W. H., and TAYLOR, T. H. (1952). Ecological and physiological factors in compounding forage seed mixtures. *Adv. Agron.* **4,** 179–221.
MASEFIELD, G. B. (1957). The value of legumes in crop associations in the tropics. *Wld Crops* **9,** 479–81.
WAGNER, R. E. and WILKINS, H. L. (1947). The effect of legumes on the percentage of crude protein in orchard grass and brome grass. *J. Am. Soc. Agron.* **39,** 141–5.
WALKER, T. W. *et al.* (1954). The nitrogen economy of grass-legume associations. *J. Br. Grassld Soc.* **9,** 249–74.

Chapter 9

Text references

NYE, P. H. (1952). *Emp. J. exp. Agric.* **20,** 227–33.
PATERSON, D. D. (1938). *Emp. J. exp. Agric.* **6,** 323–40.

Suggested further reading

COMMONWEALTH BUREAU OF PASTURES AND FIELD CROPS (1945). The establishment and early management of sown pastures. *Bull. Commonw. Bur. Past. Fld Crops* **34.**

Chapter 10

Text references

ARONOVICH, S., SERPA, A., and RIBEIRO, H. (1970). *Proc. 11th Int. Grassld Congr.* 796.
COLMAN, R. L. (1968). *Proc. Aust. Grassld Congr. (Perth)* 97.
de GEUS, J. and t'HART, M. L. (1952). *Proc. 6th Int. Grassld Congr.* 371–6.
DICKSON, G. R. (1959). *J. Br. Grassld Soc.* **14,** 1726.
DUNWELL, G. H. (1967). *Qd. agric. J.* **93,** 577.
EVANS, T. R. (1970). *Proc. 11th Int. Grassld Congr.* 803.
FREER, M. (1959). *J. agric. Sci., Camb.* **52,** 129–36.
HODGSON, J. and SPEDDING, C. R. W. (1966). *J. agric. Sci., Camb.* **67,** 150.
HOLMES, W. (1952). *Proc. 6th Int. Grassld Congr.* 377–83.
HUNT, I. V. (1952). *Scott. J. Agric.* **32,** 92–6.
HUTTON, E. M. (1968). *Proc. Aust. Grassld Congr.* 49–55.
JONES, M. G. (1933). *Emp. J. exp. Agric.* **1,** part 1, 43–57; part 2, 122–8; part 3, 223–34; part 4, 361–8.
KEMP, A. (1958). *Neth. J. agric. Sci.* **6,** 281.
LARGE, R. V. and SPEDDING, C. R. W. (1966), *J. agric. Sci., Camb.* **67,** 41.
LEHR, J. J. (1951). *Soil Sci.* **72,** 157.
LEVY, E. B. (1955). *Grasslands of New Zealand.* Whitcombe and Tombs, Wellington.
MCMEEKAN, C. P. (1960). *Proc. 8th Int. Grassld Congr.* 21.
—— and HANCOCK, J. (1954). *Proc. N.Z. Soc. Anim. Prod.* **14,** 111–17.
MITCHELL, K. J. and COLES, S. T. C. (1955). *N.Z. Jl Sci. Technol.* **A36,** 586.
NOWAKOWSKI, T. Z. (1958). *Rep. Rothamsted exp. Stn 1958* 48.
PEART, J. R. (1968). *Proc. Aust. Soc. Anim. Prod.* **7,** 110.
RAINS, A. B. and FOSTER, W. H. (1956–7). *Rep. Dep. Agric. nth. Reg. Nigeria* **2,** 163–4.
RAUZI, F. and LANG, R. L. (1956). *Bull. Wyo. agric. Exp. Stn* **344,** 1–11.
SALETTE, J. E. (1970). *Proc. 11th Int. Grassld Congr.* 404.
SEARS, P. D. (1950). *J. Br. Grassld Soc.* **5,** 267–80.
—— (1953). *N.Z. Jl Sci. Technol.* **35,** Sec. A, Suppl. 1, 1–29.
VOISIN, A. (1959). *Grass productivity.* Crosby Lockwood, London.

WALLACE, L. R. (1955). *Proc. N.Z. Grassld Ass.* 106–18.

WATSON, S. J. and RUNCIE, K. V. (1960). *Outl. Agric.* **2,** 264.

Suggested further reading

ALDER, F. E. (1954). Some effects of fertilizers on the output of leys with particular reference to extension of the grazing season. *J. Br. Grassld Soc.* **9,** 29–33.

—— (1955). Grassland management for meat production. *J. Br. Grassld Soc.* **10,** 115–25.

ARNOLD, G. W. and HOLMES, W. (1958). Studies in grazing management: VII, The influence of strip grazing versus controlled free grazing on milk yield, milk composition and pasture utilization. *J. agric. Sci., Camb.* **51,** 248–56.

BROUGHAM, R. W. (1956). Effect of intensity of defoliation on re-growth of pasture. *Aust. J. agric. Res.* **7,** 377–87.

BRUNDAGE, L. and PETERSON, W. E. (1952). A comparison between daily rotational grazing and continuous grazing. *J. Dairy Sci.* **35,** 623–30.

EDWARDS, D. C. (1940). The reaction of Kikuyu grass herbage to management. *Emp. J. exp. Agric.* **8,** 9.

HALL, T. D., MEREDITH, D., and ALTONA, R. E. (1950). Production from grassland in South Africa: fertilizer treatments and livestock gains on veld. *Emp. J. exp. Agric.* **18,** 8–18.

HOLMES, W. (1954). Modern methods of grassland management in feeding the dairy herd. *Proc. Nutr. Soc.* **13,** 19–22.

—— WAITE, R., FERGUSON, D. L., and CAMPBELL, JEAN I. (1950, 1952). Studies in grazing management: I, A comparison of production obtained from close-folding and rotational grazing dairy cows; IV, A comparison of close-folding and rotational folding of dairy cows on intensively fertilized pasture. *J. agric. Sci., Camb.* **40,** 381–91; **42,** 304–13.

IGNATIEFF, V. and PAGE, H. J., eds. (1958). *The efficient use of fertilizers* (Studies in Agriculture no. 43). 2nd edn, F.A.O., Rome.

JOHNSTONE-WALLACE, D. B. (1937). The influence of grazing management and plant associations on the chemical composition of pasture plants. *J. Am. Soc. Agron.* **29,** 441–55.

JONES, I. I. (1933). The effect of varying the period of rest on rotational grazing. *Welsh J. Agric.* **9,** 159–70.

MILTON, W. E. J. and DAVIES, R. O. (1947). The yield, botanical and chemical composition of Natural Hill Herbage under manuring, controlled grazing and hay conditions: Part 1, botanical section; Part II, chemical section. *J. Ecol.* **35,** 65–95.

UNITED KINGDOM MINISTRY OF AGRICULTURE (1944, reprinted 1952). Rotations. *Bull. Minist. Agric. Fish.* **85**.

UNITED KINGDOM MINISTRY OF AGRICULTURE (1952) Grassland management. *Bull. Minist. Agric. Fish.* **154.**

MOTTA, M. S. (1956). Grazing management and production of some tropical grasses in the British West Indies. *Proc. 7th Int. Grassld Congr.* 539.

SAMPSON, A. W. (1952). *Range management, principles and practice.* Chapman and Hall, London.

STODDART, L. A. and SMITH, A. D. (1943). *Range management.* McGraw-Hill, New York.

STRANGE, R. (1958). Preliminary trials of grasses and legumes under grazing. *E. Afr. agric. J.* **24,** 92–102.

WAITE, R., HOLMES, W., CAMPBELL, JEAN I., and FERGUSON, D. L. (1950). Studies in grazing management: II, The amount and chemical composition of herbage eaten by dairy cattle under close-folding and rotational methods of grazing. *J. agric. Sci., Camb.* **40,** 392–402.

—— MACDONALD, W. B., and HOLMES, W. (1951). Studies in grazing management: III, The behaviour of dairy cows grazed under close-folding and rational systems of management. *J. agric. Sci., Camb.* **41,** 163–73.

Chapter 11

Text references

AHLGREN, G. H. (1959). I.C.A. Project 61–13–050.

HAMILTON, R. A. (1956). Utilization of grassland. Outl. Agric. **1,** 7.

HEDRICK, D. W. (1960). I.C.A. Project 61–30–050.

MCILROY, R. J. (1962). Grassland improvement and utilization in Nigeria. *Outl. Agric.* **3,** 174–9.

MILLER, T. B. (1960). *W. Afr. J. biol. Chem.* **4,** 29–35.

OAKES, A. J. (1960). *Proc. 8th Int. Grassld Congr.* 386–9.

OGOR, E. and HEDRICK, D. W. (1960). Management of natural pasturage in Western Nigeria (Paper read to the West African Science Association Conference, Ibadan, December, 1960).

OGOR, E. (1961). Development and carrying capacity of pasture in the derived savannah of the Western Region of Nigeria. *Bull. agric. Dep. Nigeria.*

OYENUGA, V. A. (1958). *Nutr. Abstr. Rev.* **28,** 985–1000.

WHYTE, R. O. (1960). *8th Internat. Grassld Congr.* plenary lecture.

Suggested further reading

DAVIES, W. (1951). The production and utilization of grass throughout the year. *Proc. Br. Soc. Anim. Prod.* 73–89.

DELHAYE, R. E. (1959). How to manage and improve pastures in the lower Congo (Mouazi). *Bull. Inf. I.N.E.A.C.* **8,** 35–49.

GERMAIN, R. and SCAUT, A. (1960). Herbage and nutritional aspects of animal husbandry in the equatorial forest of the Congo. *Proc. 8th Int. Grassld Congr.* 371.

HOLMES, W. (1954). High milk yields per acre from grassland. *J. Br. Grassld Soc.* **9,** 17–27.

—— (1962). Grazing management for dairy cattle. *J. Br. Grassld Soc.* **17,** 30.

MICHEL, G. (1958). Trials in stocking pastures at Rubona (Congo) in 1953–6. *Bull. Inf. I.N.E.A.C.* **7,** 229–36.

INSTITUT NATIONAL POUR L'ETUDE AGRONOMIQUE DU CONGO BELGE (1958, 1959, 1960). Utilization of sown pastures at Yangambi. *Reports* **90, 102, 211.**

Chapter 12

Text references

COMPERE, R. (1961). Productivity and feeding value of the forage of some pasture types studied at Mulungu (Kivu). *Bull. agric. Congo belge* **52,** 61–93.

DABADGHAO, P. M. (1960). *Proc. 8th Int. Grassld Congr.* 226.

—— and SHANKARNARAYEN, K. A. (1970). *Proc. 11th Int. Grassld Congr.* 36.

DAVIES, J. G. (1965). *Proc. 9th Int. Grassld Congr.* **2,** 11.

DELHAYE, R. E. (1959). How to manage and improve pastures in the lower Congo (Mouazi). *Bull. Inf. I.N.E.A.C.* **8,** 35–49.

FORSTER, R. H., WILSON, P. N., and BUTTERWORTH, M. H. (1960). *Proc. 8th Int. Grassld Congr.* 390.

FOSTER, W. H. (1961). *Emp. J. exp. Agric.* **29,** 319.

GROSSMAN, J. (1956). *Proc. 7th Int. Grassld Congr.* 528.

INSTITUT NATIONALE POUR L'ETUDE AGRONOMIQUE DU CONGO BELGE (1958, 1959, 1960). *Reports.*

KATHERINE RESEARCH STATION (1959) *Progress Report* 1954–6 C.S.I.R.O., Melbourne.

MCILROY, R. J. (1962). Grassland improvement and utilization in Nigeria. *Outl. Agric.* **3,** 174–9.

MOTTA, M. S. (1956). *Proc. 7th Int. Grassld Congr.* 539.

ROCKEFELLER FOUNDATION (1960–1). *Agricultural science annual report.*
SHAW, N. H. (1961). *Aust. J. exp. Agric. Anim. Husb.* **1**, 73.
VOISIN, A. (1960). *Better grassland sward.* Crosby Lockwood, London.
WHYTE, R. O. and KUMAR, L. S. S. (1957). *Bull. Indian Coun. agric. Res.* **31.**

Suggested further reading

SEMPLE, A. T. (1952). *Improving the world's grasslands.* F.A.O., Rome.
WHYTE, R. O. (1968). *Land, livestock and human nutrition in India.* Praeger, New York, Washington, and London.

Chapter 13

Text references

BOYNS, B. M. (1955). *Studies in agriculture* no. 42, 146. F.A.O., Rome.
DUCKWORTH, J. (1949). *Trop. Agric., Trin.* **26**, 24–6.
FRENCH, M. H. (1941). *Emp. J. exp. Agric.* **9**, 23–8.
MINSON, D. J. and BROWN, Miss S. A. (1959). Grassland Symposium, Soc. Chem. and Ind., Dublin.
MINSON, D. J., TAYLER, J. C., ALDER, F. E., RAYMOND, W. F., and RUDMAN, J. E. (1960). *J. Br. Grassld Soc.* **15**, 86–8.
NORDFELT, S. *et al.* (1951). *Bull. Hawaii agric. Exp. Stn* **12**.
OYENUGA, V. A. (1957). *Emp. J. exp. Agric.* **15**, 242.
—— (1959). *W. Afr. J. biol. Chem.* **3**, 43–58.
—— (1960*a*). *J. agric. Sci., Camb.* **55**, 339–50.
—— (1960*b*). *W. Afr. J. biol. Chem.* **4**, 46–63.
PATERSON, D. D. (1933). *J. agric. Sci., Camb.* **23**, 615–41.
—— (1935). *J. agric. Sci., Camb.* **25**, 369–96.
RAINS, A. B. and FOSTER, W. H. (1956–7). *Rep. Dep. Agric. nth. Reg. Nigeria* **2**, 163–4.
RIPPERTON, J. C. and TAKAHASHI, M. (1948). *Rep. Hawaii agric. Exp. Stn.*
SCHOFIELD, J. L. (1945). *Qd J. agric. Sci.* **2**, 209–43.
WAITE, R. and SASTRY, K. N. S. (1949). *Emp. J. exp. Agric.* **17**, 179–87.
WALSH, S. R. (1958). *Qd agric. J.* **84**, 527–36.
WEINMANN, H. (1955). *Grasses and pastures of South Africa*, pp. 571–600. Central News Agency, Johannesburg.

Suggested further reading

CARTMILL, W. J. (1944). The effects of season, stage of growth and soil type on the chemical composition of grasses in the Queensland 'wet belt'. *Qd J. agric. Sci.* **1**, 1–31.

COMPERE, R. (1961). Productivity and feeding value of the forage of some pasture types studied at Mulungu (Kivu). *Bull. agric. Congo belge* **52,** 61–93.

EVANS, R. E. (1960). Rations for livestock. *Bull. Minist. Agric. Fish.* **48.**

HOLMES, W. (1952). The feeding value of grass and grassland products. *Proc. Br. Soc. Anim. Prod.* 90–112.

MINISTRY OF AGRICULTURE (1942). Composition and nutritive value of feeding stuffs. *Bull. Minist. Agric. Fish.* **124.**

SCAUT, A. (1959). Determination of the digestibility of fresh herbage (Congo). *I.N.E.A.C. Ser. Sci.,* 81–6.

SULLIVAN, J. T. and GERBER, R. J. (1947). Chemical composition of pasture plants. *Bull. Pa agric. Exp. Stn* **489.**

WATSON, S. J. and HORTON, E. A. (1936). Composition, digestibility and nutritive value of samples of grassland products. *J. agric. Sci.,* **26,** 142–55.

—— PROCTER, J., and FERGUSSON, W. W. (1932). The effect of nitrogen on the yield, chemical composition, and digestibility of grassland herbage. *J. agric. Sci., Camb.* **22,** 251–90.

WILLIAMS, E. D. (1953). Trace elements and grassland. *N.A.A.S. Camb. q. Rev.* **41.**

WOODMAN, H. E. and UNDERWOOD, E. J. (1932). The influence of intensive fertilizing on the yield and composition of good permanent pasture. *J. agric. Sci., Camb.* **22,** 26–71.

WOODMAN, H. E. and EVANS, R. E. (1938). The influence of cutting at monthly intervals over nine seasons on the quality and productivity of a heavy-land pasture. *J. agric. Sci., Camb.* **28,** 581–91.

Chapter 14

BREIREM, K. and ULVESLI, O. (1960). *Herb. Abstr.* **30,** 301.

Chapter 15

Text references

BARKER, A. S. *et al.* (1955). *J. Br. Grassld Soc.* **10,** 67.

BROWN, Miss D. (1954). Methods of surveying and measuring vegetation. *Bull. Commonw. Bur. Past. Fld Crops* **4.**

CASTLE, M. E. (1955). *Herb. Abstr.* **25,** 1–9.

CORBETT, J. L. and GREENHALGH, J. F. D. (1959). Grassland Symposium, Soc. Chem. and Ind., Dublin.

COWLISHAW, S. (1951). *J. Br. Grassld Soc.* **6,** 179–82.

DIJKSTRA, N. D. (1954). *Neth. J. agric. Sci.* **2,** 273.

IVINS, J. D., ed. (1959). *Measurement of grassland productivity*. Butterworth, London.
LANCASTER, R. J. (1949). *Nature, Lond.* **163,** 330.
LEVY, E. B. and MADDEN, E. A. (1933). *N.Z. Jl Agric.* **46,** 267.
MINSON, D. J. and RAYMOND, W. F. (1958). *Exps Prog. Grassld Res. Inst.* **10,** 92.
MINSON, D. J. and BROWN, Miss S. A. (1959). *Exps Prog. Grassld Res. Inst.* **11,** 99.
MINSON, D. J. *et al.* (1960). *J. Br. Grassld Soc.* **15,** 86.
MACLUSKY, D. S. (1955). *Proc. Br. Soc. Anim. Prod.* 45.
RAYMOND, W. F. (1948). *Nature, Lond.* **161,** 937.
—— (1954). *J. Br. Grassld Soc.* **9,** 61–7, 69–84.
—— and MINSON, D. J. (1955). *J. Br. Grassld Soc.* **10,** 282.
—— *et al.* (1959). *Measurement of grassland productivity* (ed. J. D. Ivins), pp. 156–64. Butterworth, London.
SAMPFORD, M. R. (1960). *Proc. 8th Int. Grassld Congr.* 742.

Suggested further reading

CHRISTIAN, K. R. and COOP, M. R. (1954). *N.Z. Jl Sci. Technol.* **36,** 328–30.

Chapter 16

Text references

BALCH, C. C. (1960). *Proc. 8th Int. Grassld Congr. (U.K.)* 528.
—— and CAMPLING, R. C. (1962). *Nutr. Abstr. Rev.* **32,** 669.
BARNES, R. F. (1965). *Agron. J.* **57,** 213.
BLAXTER, K. L. (1956). *Br. J. Nutr.* **10,** 69.
—— (1960). *Proc. 8th Int. Grassld Congr. (U.K.)* 479.
BREDON, R. M., TORRELL, D. T., and MARSHALL, B. (1967). *J. Range Mgmt* **20,** 317.
CAMPLING, R. C., FREER, M., and BALCH, C. C. (1962). *Br. J. Nutr.* **16,** 115.
CRAMPTON, E. W., DONEFER, E., and LLOYD, L. E. (1960*a*). *Proc. 8th Int. Grassld Congr. (U.K.)* 462.
—— (1960*b*). *J. Anim. Sci.* **19,** 538, 545.
DENT, J. W. (1963). Proc. 1962 Crop Conf. N.I.A.B. Cambridge.
GALLAGHER, J. R. and WATKIN, B. R. (1967). *J. agric. Sci., Camb.* **68,** 87.
GLOVER, J. and DOUGALL, H. W. (1960). *J. agric. Sci., Camb.* **55,** 391.
GREENHALGH, J. F. D. and REID, G. W. (1967). *Nature, Lond.* **214,** 744.
GRIMES, R. C., WATKIN, B. R., and GALLAGHER, J. R. (1967). *J. agric. Sci., Camb.* **68,** 11.

HARKESS, R. D. (1963). *J. Br. Grassld Soc.* **18,** 62.
HI KON OH, BAUMGARDT, B. R., and SCHOLL, J. M. (1966). *J. Dairy Sci.* **49,** 850.
HOLDER, (1966). *Rep. Grassld Res. Inst.*
HUTTON, J. B. (1962). *N.Z. Jl agric. Res.* **5,** 409.
INGALLS, J. P., THOMAS, J. W., BENNE, E. J., and TESAR, M. (1965). *J. Anim. Sci.* **24,** 1159.
JARRIGE, R., THE VEND, P., DEMARQUILLY, C. (1970). *Proc. 11th Int. Grassld Congr.* 762.
JOHNS, A. T. (1965). *Proc. 9th Int. Grassld Congr.* (*Brazil*) 9; 145.
JOHNSON, P. R., DEHORITY, R. A., and PARSONS, J. W. (1965). *Proc. 9th Int. Grassld Congr.* (*Brazil*) 9; i.p.; 587.
KINGSBURY, L. A. (1965). *Proc. N.Z. Soc. Anim. Prod.* 119.
LAMBOURNE, L. J. (1964). *Exps Prog. Grassld Res. Inst.* **16,** 68.
LLOYD, L. E. *et al.* (1961). *J. Anim. Sci.* **20,** 468.
MCLEOD, M. N. and MINSON, D. J. (1970). *J. Br. Grassld Soc.* **25,** 296.
MILFORD, R. and MINSON, D. J. (1965*a*). *Proc. 9th Int. Grassld Congr.* (*Brazil*) 9, i.p. 236.
—— (1965*b*). *Herb. Abstr.* **35,** 1830.
—— (1966). Ann. Rpt. Trop. Pastures C.S.I.R.O.
MINSON, D. J. (1968). *Proc. Aust. Grassld Congr.* (*Perth*) 25.
—— and MCLEOD, M. N. (1970). *Proc. 11th Int. Grassld Congr.* 719.
—— *et al.* (1960). *J. Br. Grassld Soc.* **15,** 174.
—— HARRIS, C. E., RAYMOND, W. F., and MILFORD, R. (1964). *J. Br. Grassld Soc.* **19,** 298.
OSBOURN, D. F., THOMSON, D. J., and TERRY, R. A. (1966). *Proc. 10th Int. Grassld Congr.* (*Helsinki*) 363.
REID, R. L. and JUNG, G. A. (1965). *Proc. 9th Int. Grassld Congr.* (*Brazil*) **9,** 418.
——, —— and MURRAY, S. J. (1966). *J. Anim. Sci.* **25,** 636.
ROOK, J. A. F. (1959). *Proc. Nutr. Soc.* **18,** 117.
TILLEY, J. M. A. and TERRY, R. A. (1963). *J. Br. Grassld Soc.* **18,** 104.
——, DERIAZ, R. E., and TERRY, R. A. (1960). *Proc. 8th Int. Grassld Congr.* (*U.K.*) 533.
VAN SOEST, P. J. (1965). *J. Anim. Sci.* **24,** 834.
WEIR, W. C. and TORRELL, D. T. (1959). *J. Anim. Sci.* **18,** 641.
——, MEYER, J. H., and LOFGREEN, G. P. (1959). *Agron. J.* **51,** 235.

Appendix B

MARTIN, W. S. (1944). *E. Afr. agric. J.* **9,** 189–95.

Index